WITHDRAWN
FROM
UNIVERSITIES
AT
MEDWAY
LIBRARY

Current Topics in Vector Research

Current Topics in Vector Research

Edited by

Kerry F. Harris
Virus–Vector Laboratory, Department of Entomology, Texas A&M University, College Station, Texas 77843, USA

Editorial Board

Current Topics in Vector Research

Volume 4

Edited by Kerry F. Harris

With Contributions by
M.E. Coan J.E. Duffus B.F. Eldridge
F.E. Gildow A.L. Hicks J. Kramer
P.G. Lawyer D.W. Tonkyn
R.F. Whitcomb D.G. Young

With 51 Illustrations

Springer-Verlag
New York Berlin Heidelberg
London Paris Tokyo

Kerry F. Harris
Virus-Vector Laboratory
Department of Entomology
Texas A&M University
College Station, Texas 77843
USA

Volumes 1 and 2 of *Current Topics in Vector Research* were published by Praeger Publishers, New York, New York.

ISSN: 0737-8491

Typeset by David E. Seham Associates Inc., Metuchen, New Jersey.
Printed and bound by R.R. Donnelley and Sons, Harrisonburg, Virginia.
Printed in the United States of America.

9 8 7 6 5 4 3 2 1

ISBN 0-387-96464-9 Springer-Verlag New York Berlin Heidelberg
ISBN 3-540-96464-9 Springer-Verlag Berlin Heidelberg New York

Preface

Current Topics in Vector Research is based on the premise that to understand the whole, one must first understand the component parts and how they interact. Here in Volume 4, as well as in future volumes, vector, pathogen, and host will be treated both individually and as integral parts of multifaceted transmission systems. It is our intention to present up-to-date, coherent syntheses of the latest findings in vector research, suggest promising frontiers for future research, and call attention to possible practical applications of our present understandings of pathogen-vector-host interactions. To realize our goals, we invite world-renowned, veteran scientists as well as neophytes to report on their individual areas of expertise. Where appropriate, authors are encouraged to draw conclusions and propose hypotheses that stimulate additional thinking and research or otherwise further our understanding of vector transmission cycles and how such cycles might be interrupted. It is our hope that readers will agree that we are serving these objectives and creating a milieu for specialists and generalists in vector research to maintain rapport and understanding.

I wish to thank the members of the Editorial Board for their encouragement and help in bringing Volume 4 of *Current Topics in Vector Research* to fruition. We especially thank the authors in Volume 4 for their outstanding contributions. Each contributor was challenged to be as creative as possible in presenting their analyses of data, syntheses of ideas, and writing in order to point out where we might or should be headed in their areas of vector research, not just where we have been. Without exception, they responded to this challenge and addressed their topics in thorough and imaginative manners.

Bruce Eldridge gets us off to a fine start in Chapter 1 with his article on the means by which mosquito-borne arboviruses survive seasons unfavorable for active mosquito transmission—a question that has perplexed biologists for decades. Understanding of a possible role of females of species of the genus *Culex* as overwinter hosts for arboviruses for which they serve as vectors is clouded by our lack of understanding of the complex phenomenon of diapause in these mosquitoes. Eldridge presents us with a comprehensive review of diapause in *Culex* mosquitoes in the context of arbovirus ecology. New data are included, and new research directions are suggested, which could shed light on this important area of medical entomology. In Chapter 2, David Young and Phillip Lawyer emphasize the great diversity of *Leishmania* parasites and their *Lutzomyia* vectors in the Americas. Recent advances in identifying *Leishmania* by biochemical methods have provided a basis for incriminating vector species in specific foci. The authors discuss and summarize existing knowledge (or the lack of it) of the transmission cycles in these foci and the biology and vector competence of sand flies. Beginning with Chapter 3 and continuing through Chapter 6, we change gears with four chapters aimed at bettering our understanding of pathogen-vector-host plant interactions.

Whitefly-transmitted viruses cause significant losses in agriculture throughout the world, especially in tropical and subtropical areas, but in recent years, also in wide areas north and south of the tropics, approaching areas of intensive agricultural production such as the southern United States, Jordan, and Israel. In Chapter 3, James Duffus summarizes our knowledge, much of it quite recent, of these vectors and the viruses that they transmit. A compilation of available data suggests at least seven groups of whitefly-transmitted viruses, plus an eighth group with unknown relationships, based on such characteristics as particle morphology, serological relationships, vector relationships, types of transmission persistence (nonpersistent, semipersistent, or persistent), transstadiality (whether or not vector inoculativity is retained through molts), symptom expression, and mechanically transmissible or not. Individual viruses can be categorized according to their relatedness or possible relatedness to one of several virus groupings: those resembling geminiviruses, carlaviruses, closteroviruses, potyviruses, luteoviruses, DNA-containing rod-shaped viruses, or nepoviruses, and those of unknown relatedness. Duffus makes it readily apparent that whitefly-transmitted viruses are much more diverse than originally thought with respect to mode of transmission and particle morphology. He presents the kinds of information that will prove invaluable in the formulation of disease control strategies, especially those involving modification of cultural practices.

Chapter 4 provides us with an up-to-date description of recent research directed at better understanding persistent transmission of plant viruses by aphids. Its author, Frederick Gildow, uses the plant virus-aphid membrane interactions involved in the transmission of luteoviruses to describe a model system for explaining the cellular and molecular basis of vector specificity. Results from a variety of ultrastructural studies and transmission bioassay tests are described, which confirm the hypothesis that luteoviruses are internalized into the aphid gut and salivary gland epithelium by receptor-mediated endocytosis, transported across these cells in membrane-bound organelles, and subsequently released for transmission by exocytosis. A discussion of the contributions from several laboratories leading to our present state of knowledge concerning persistent transmission is given, as well as suggestions for future directions of this research.

In the fifth chapter, Robert Whitcomb, James Kramer, Michael Coan, and Andrew Hicks summarize accumulated observations concerning grassland cicadellids and their hosts. In so doing, they tie together work that spans nearly a century. Catalyzed by the evolutionary perspectives of the late H.H. Ross, Whitcomb and his colleagues emphasize evolutionary, biogeographic, and biosystematic aspects of the involved cicadellid genera. At a time when the majority of ecologists are examining proximate factors that regulate populations in very limited geographic areas, this treatment offers a refreshingly important overview. Eventually, synthesis of detailed analyses with the broad regional perspectives of Whitcomb and his associates will permit a holistic understanding of cicadellid ecology, and with it, answers to the puzzling question of pest appearance.

In the final chapter in this volume, Chapter 6, David Tonkyn and Robert Whitcomb review the literature on Homoptera feeding to make the following five points regarding feeding strategies: (1) Homoptera typically feed either on phloem or xylem sap or on the cytoplasm of mesophyll and other cells, though there are exceptions that feed more generally. (2) Each of these three feeding modes requires special morphologic, physiologic, and behavioral adaptations, which may preclude feeding on other plant, let alone animal, tissues. (3) As a result, the food eaten by these insects is a conservative character in evolution, and often characterizes entire subfamilies or families. (4) Other features of the biology of the Homoptera, such as patterns of host-specificity and their role as vectors of plant pathogens, may also be understood in light of their feeding strategies. On the basis of the aforementioned four points, it is useful to categorize, where possible, Homoptera into three distinct feeding guilds. (5) Plant pathogens, such as bacteria, fungi, viruses, and spiroplasmas, that inhabit the vascular tissues also tend to be tissue specialists, though they can subdivide these tissues on a much finer spatial and temporal scale. As a result, the pathogens and the Homoptera that transmit them are best considered members of different, only partially overlapping guilds.

We thank the staff of Springer-Verlag for their encouragement and technical assistance.

Kerry F. Harris

Contents

Contributors

Michael E. Coan
Hemoparasitic Disease Research Unit, Agricultural Research Service, United States Department of Agriculture, Pullman, Washington 99164

James E. Duffus
Agricultural Research Service, United States Department of Agriculture, 1636 East Alisal Street, Salinas, California 93905

Bruce F. Eldridge
Department of Entomology, University of California, Davis, California 95616

Frederick E. Gildow
Department of Plant Pathology, The Pennsylvania State University, University Park, Pennsylvania 16802

Andrew L. Hicks
Insect Pathology Laboratory, Agricultural Research Service, United States Department of Agriculture, Beltsville, Maryland 20705

James Kramer
Systematic Entomology Laboratory, Agricultural Research Service, United States Department of Agriculture, Washington, D.C. 20036

Phillip G. Lawyer
Department of Entomology, Walter Reed Army Institute of Research, Walter Reed Army Medical Center, Washington, D.C. 20307

David W. Tonkyn
Department of Biological Sciences, Clemson University, Clemson, South Carolina 29634

Robert F. Whitcomb
Insect Pathology Laboratory, Agricultural Research Service, United States Department of Agriculture, Beltsville, Maryland 20705

David G. Young
Department of Entomology and Nematology, University of Florida, Gainesville, Florida 32611

1
Diapause and Related Phenomena in *Culex* Mosquitoes: Their Relation to Arbovirus Disease Ecology

Bruce F. Eldridge

Introduction

Most mosquitoes in the genus *Culex* appear to overwinter as inseminated female adults, although individuals that have actually been observed overwintering in nature represent only a few species. Moreover, searches for such relatively common species as *Culex salinarius* Coquillett and *C. restuans* Theobald have repeatedly been unsuccessful. Aside from the importance of mosquito diapause as a biological phenomenon, there has been considerable interest in the subject because of the possibility that overwintering female mosquitoes may serve as hibernal reservoirs of arboviruses causing human and animal diseases (33, 81). Two North American mosquito-borne arboviruses of public health importance are transmitted by species of the genus *Culex:* St. Louis encephalitis (SLE) by *C. pipiens* Linneaus and *C. tarsalis* Coquillett, and western equine encephalomyelitis (WEE) transmitted by *C. tarsalis*. In Asia, *C. tritaeniorhynchus* Giles is the vector of Japanese encephalitis (JE) in temperate regions. Until the convincing demonstration of transovarial transmission of an arbovirus by a species of mosquito (115), few biomedical scientists gave much credence to the idea that mosquitoes played an important role as winter reservoirs of arboviruses. The isolation of SLE virus from overwintering *Culex pipiens* (3) focused attention on *Culex* mosquitoes in this context, although these isolations were not novel. Isolations from overwintering *Culex* mosquitoes had been reported of WEE (14, 82) and of JE (43, 58, 111). Reeves *et al.* (82) reported a probable isolation of SLE virus from a pool of *C. quinquefasciatus* Say collected in November in Kern County, California. Although the general characteristics of overwintering in *Culex* were described 50–70 years ago using anautogenous *C. pipiens* as a model (15–17, 22, 45, 46, 56, 71, 80, 88, 89, 109), many

Bruce F. Eldridge, Department of Entomology, University of California, Davis, California 95616, USA.

ecological and physiological details of the phenomenon are still poorly understood.

The purpose of this chapter is to review the current status of knowledge of diapause in *Culex* mosquitoes, present some previously unpublished information, and review information on the related topics of autogeny and transovarial transmission of arbovirus pathogens. Finally, I will discuss these phenomena in relation to the question of mechanisms of overwinter survival of arboviruses. Most of the information discussed will pertain to species classified in the subgenus *Culex* of the genus *Culex*. Although our knowledge of diapause ecology is fragmentary for members of this subgenus, it is virtually nonexistent for members of other *Culex* subgenera. These subgenera are poorly represented in Holarctica, however, and are probably of little importance as disease pathogen vectors in colder regions.

A caveat appears in order: we are at the dawn of many significant discoveries in the area of arbovirus disease ecology. This new horizon derives from the present availability of powerful techniques which will enhance our ability to detect virus particles in biological systems at a high level of sensitivity, and to observe directly other phenomena with relative ease. Accordingly, the following fund of information is in a state of flux. The developments of the past decade will profoundly influence those of the next.

Terminology and Characteristics of Diapause

Over the years a number of terms have been applied to phenomena in *Culex* mosquitoes related to overwintering. The term diapause itself has been used in various contexts. I use the term here to refer to a physiological state of an individual mosquito that prepares the individual for cold winter conditions, thereby maintaining the organism during winter. The state is generally triggered by some factor other than cold temperature alone. This definition is consistent with the definition of "diapause" of Mansingh (63). Although diapause can arise in response to any adverse conditions, use of the term here should be understood to mean winter diapause. Overwintering is a general term, which refers to any means of surviving winter and is considered synonymous with hibernation.

Diapause in *Culex* mosquitoes is characterized by (1) ovarian diapause, (2) reduced blood avidity, (3) hypertrophy of the fat body, and (4) relative inactivity. At emergence, ovarioles in *Culex* mosquitoes consist of a germarium and an undifferentiated follicle of about equal size [Stage No. 2 of Kawai (51)]. In gonoactive females, the follicle grows within several days (the time is temperature-dependent) to the resting stage. This represents the completion previtellogenic development. The follicles will grow no further in anautogenous females unless a blood meal is taken. The stage of development at which the resting stage occurs may vary among species. In most species studied, suspension of growth occurs when the primary follicle becomes about two and one-half times as large as the germarium; the oocyte becomes differentiated from the surrounding nurse cells and yolk fills between one-fourth and one-half of the oocyte [Stage IIb of Kawai (51)]. In *Aedes aegypti,* scattered microvilli, pits, and pinocytotic vesicles are present on the surface of the oocyte at this stage (2).

Ovarian diapause is a condition in which the ovarian follicles of the female mosquito do not develop past Stage N of Kawai (51) or a follicle:germarium ratio of about 1.5:1. (99, 105). For *C. pipiens,* diapausing follicles measure no more than 0.05 mm; nondiapausing (resting stage) follicles measure 0.07 mm or more (34). When ovaries are in the diapause state, there is reduced blood avidity, but if a blood meal is taken, the ovaries remain undeveloped. The term gonotrophic dissociation has been applied to this situation in *Culex* mosquitoes (28, 32, 75, 77), but was referred to as "none" in the review of terms by Eldridge (33). The phenomenon "gonotrophic dissociation" comprises both physiological and behavioral factors and was coined to designate certain anophelines that take blood periodically during the winter without subsequent ovarian development (114). It should not be applied to *Culex* mosquitoes, because even a prehibernation blood meal is probably rare by diapausing females, let alone repeated blood meals during hibernation. The term "potential gonotrophic dissociation" may be appropriate in the sense that the term "potential diapause" has been used for nondiapausing species such as *Aedes aegypti* (Linneaus) (104).

Methods

Experiments described in this chapter were done with natural and colonized strains of *C. pipiens, C. peus* Speiser, and *C. tarsalis* from Benton County, Oregon and with colonized strains of *C. tarsalis* from Kern and Yolo Counties, California. Collections of larvae of all three species were made from a single log pond in Benton County, Oregon. Winter collections of *C. pipiens* and *C. tarsalis* were made from under concrete highway bridges between Blodgett and Philomath, Benton County, Oregon. Winter collections of *C. pipiens* were also made from abandoned underground ammunition bunkers at Camp Adair, Benton County, Oregon. Adult female mosquitoes collected in winter were lightly anesthetized, identified to species, and then frozen.

Colonized material was maintained under insectary conditions of 28°C and a photoperiod of L:D 16:8. In experimental manipulations described, laboratory-reared pupae were randomly separated into treatment groups of 100 or more and placed in incubators programmed for various conditions of temperature and photoperiod. At various times after adult emergence, subsamples of adult females were removed from treatments and frozen for later processing.

Frozen material was thawed and then either dissected to determine state of ovarian development, or dried and extracted with petroleum ether to determine amount of fat present. To determine state of ovarian development, ovaries were teased from mosquito abdomens with needles into a drop of physiological saline and "disrupted" with a dissection needle. Ten follicles were measured with the aid of a squared reticle in the eyepiece of a compound microscope at 40×, five from each ovary, to furnish a mean value. The most advanced follicles found were always selected for measurement. Ether extraction was done in a Soxhlet apparatus, ten individual mosquitoes at a time. Specimens were first dried to constant weight in an oven at 70°C. Weights were determined on an electronic balance. The difference between preextraction and postextraction weights was considered to represent extracted lipids.

Diapause Occurrence in *Culex* Species

The species of *Culex* that undergo diapause experimentally or have been collected as adult females under natural conditions during winter and presumed to be in diapause are shown in Table 1.1. *Culex quinquefasciatus, C. salinarius, C. erythrothorax* Dyar, and *C. nigripalpus* Theobald generally occupy more southern ranges in North America, and can persist as larvae during the winter if temperatures are warm enough (24, 69). Experimentally, it has been shown that *C. quinquefasciatus* (32, 118) and *C. salinarius* (35, 61, 101) do not undergo diapause. This presumably limits the northern extension of their ranges. On the other hand, Reisen found that 51% of female *C. quinquefasciatus* collected in resting shelters in the fall and winter in Kern County, California had ovaries with follicles in the diapause state (86). Likewise, Wallis *et al.* (113) report collection of

TABLE 1.1. Species of *Culex* mosquitoes for which diapause has been demonstrated under laboratory conditions (L) or that have been collected as adult females during winter under natural conditions (N).

Species	Conditions	Reference
Culex apicalis	N	13, 94
Culex peus	L	99
Culex pipiens	N	22
	L	28, 31, 105, 112
Culex restuans	L	35, 36, 61
	N	44
Culex salinarius	N	113
Culex tarsalis	N	12, 25, 65
	L	42, 66
Culex territans	N	48
Culex tritaeniorhynchus	N	20
	L	30, 77, 78

diapausing *C. salinarius* in Connecticut, but they do not mention specific dates of collection, nor criteria used to determine diapause. *Culex peus* is enigmatic in that its range extends only a little north of the Columbia River in the Pacific Northwest, yet it undergoes diapause experimentally (99). More studies with populations of *Culex* mosquitoes from different latitudes are needed, especially for species with large latitudinal ranges, such as *C. tarsalis*. Although the subject of winter biology has been addressed in detail by several workers (10, 70), there is little information available concerning latitudinal variation in diapause even for this important species.

Diapause Induction

Ecological Factors

It has been shown repeatedly that *Culex* mosquitoes that undergo diapause in temperate regions do so in response to exposure to short photophases of late summer and autumn (28, 31, 35, 36, 42, 73, 92, 99, 105). The response is nearly always the "long-day" or "type I" response (8). The photoperiod-sensitive stages are late-stage larvae and pupae (32, 84, 105). In some species, notably *C. pipiens,* this photoperiod response has been shown to be subject to modification by temperature under laboratory conditions. This "temperature threshold" will vary with the geographic source of the strain under consideration. Thus females of this species from Lafayette, Indiana (32) and Boston, Massachusetts (105) will undergo ovarian diapause in response to short photophases only if the photoperiod-sensitive stages and the adult stages are maintained at temperatures of 18°C or lower. I tested a colonized strain of *C. pipiens* from Benton County, Oregon, and subjected individuals to different combinations of temperature and photoperiod from the pupal stage until 25 days after emergence of adults. The results are shown in Figure 1.1. Follicular length was used as a gauge of ovarian diapause. For this strain, the temperature threshold was 21°C. The decrease in follicular length in all groups after day 10 or day 12 postemergence may be artifactual and a result of maintaining adults under unnatural combinations of temperature and photoperiod for long periods of time. Such decreases are not apparent in natural overwintering populations (Figs. 1.6 and 1.7), and females that are apparently parous (those with follicular relics) are rarely found overwintering (21, 95). Follicular atresia is known to occur in bloodfed female mosquitoes, especially in cases where the size of the blood meal is smaller than normal (27, 57), but it can also occur in nonbloodfed females. Spielman and Wong (105) found that follicles of *C. pipiens* degenerated after holding nonbloodfed females under summer photoperiod conditions at 18°C. Oda *et al.* (78) observed follicular degeneration experimentally in a high proportion of nulliparous female *C. tritaeniorhynchus* maintained under summer con-

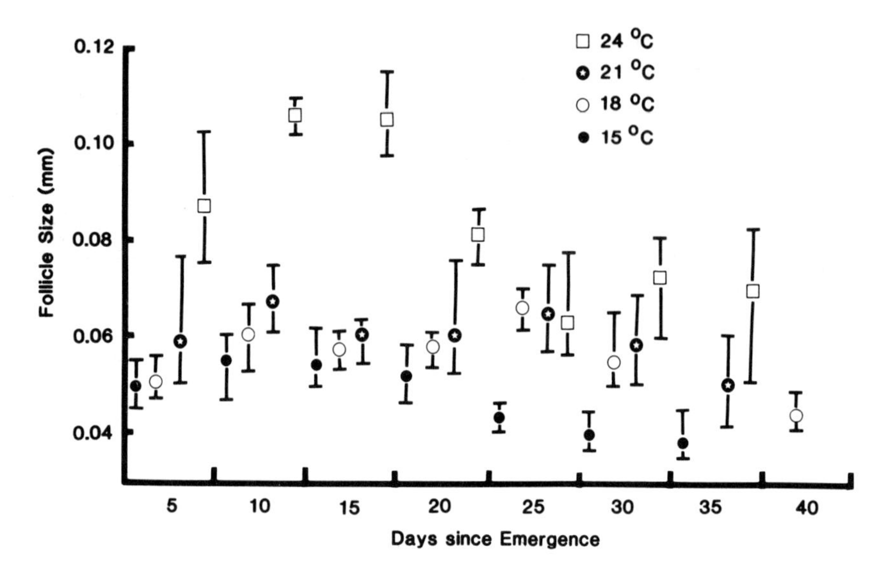

FIGURE 1.1. Ovarian follicle growth of *Culex pipiens* (western Oregon strain) under conditions of L:D 8:16 photoperiod and four different temperatures. Each symbol represents mean and range of follicle measurements from 10 females.

ditions (28°C, 16:8 L:D). When the mosquitoes were transferred to short photophase conditions at various times during their life cycles, however, the percentage of degenerating follicles was reduced. They concluded that such degeneration rarely occurred in diapausing females in nature. In experiments with *C. pipiens* from western Oregon I found that ovarian follicle size decreased gradually after 5–7 days when nonbloodfed females were held under conditions of warm temperatures (25°C) and short photophases (8:16 L:D), but not when held under warm temperatures and long photophases (16:8 L:D). After 30 days of exposure to these latter conditions, follicles were not significantly smaller in size than those examined after 5 or 10 day's exposure. More research is needed on environmental and physiological factors causing follicular atresia in both bloodfed and unfed female mosquitoes. In unfed females it probably results from levels of juvenile hormone (JH) sufficient to promote growth of follicles to the resting stage, but insufficient to maintain them at this stage. Wilson (117) has shown that JH is needed to maintain as well as promote follicular growth in *Drosophila*.

There have been many studies with other insects comparing the effect of various photoperiods on diapause induction, but few such studies with mosquitoes. Furthermore, most studies involving *Culex* mosquitoes have failed to produce the sharp classic "type I" curves as illustrated by Beck (8) for other insects, and also derived from studies of photoperiodic induction of larval diapause (e.g., 49). Typical of the graded diapause responses to a range of photoperiods are the data presented by Vinogradova (112), Sanburg and Larsen (92), and Skultab and Eldridge (99). Thus, the term "photoperiodic threshold" may have little meaning, other than in the statistical sense. Using standard statistical techniques for dosage-dependent bioassay, such as probit analysis, it is possible to obtain values for "critical photoperiods" as defined as ED_{50} values (effective "dose" of photophase producing diapause in 50% of the individuals tested), and such comparisons may be useful in comparing different populations of mosquitoes. Holzapfel and Bradshaw (47) provide a useful discussion of the statistical techniques involved.

I subjected samples of colonized strains of *C. pipiens* and *C. tarsalis* from Benton County, Oregon to a number of different photoperiods, ranging from L:D 8:16 to 18:6, all at 18°C (Fig. 1.2). Using a follicle:germarium ratio of 1.5:1 or less as a criterion of diapause and based on a minimum of ten mosquitoes at each data point, I obtained critical photophase (ED_{50}) values for *C. pipiens* of 12.71 (hours of light per 24) (95% CL 13.32 and 12.20) and for *C. tarsalis* of 14.81 (95% CL 15.81 and 14.03). Under similar experimental conditions, Skultab and Eldridge (99) reported a critical photophase for a Benton County, Oregon population of *C. peus* of 13.05 (95% CL 13.43 and 12.69). It is interesting to compare these results with routine collections of larvae from a log pond in Benton County, Oregon (Fig. 1.3). All three species occur in the same log pond. The data shown are for the period May–October 1981, but systematic larval collections over several years have produced the same general pattern. The peak of abundance of *C. tarsalis* occurs in late spring, that of *C. pipiens* in late spring to early summer, and that of *C. peus* in early fall. The seasonal disappearances of *C. tarsalis* and *C. pipiens* are generally consistent with the differences found in their critical photophase values. The seasonal abundance pattern of *C. peus* cannot be explained on the basis of critical photoperiod values, however.

There is a 6- to 8-week period between the time *C. tarsalis* larvae disappear from the log pond and the time adult females appear under concrete highway bridges in the area (compare Fig. 1.3 with Figs. 1.5 and 1.7); I presume that females are seeking carbohydrate food sources at this time, but I have not observed such feeding.

Endocrine Control

Much remains to be learned concerning the physiology of diapause induction in *Culex* mosquitoes. Ovarian diapause apparently results from juvenile hormone (JH) deficiency. Spielman (104) showed that allatectomized long-photophase females of *C. pipiens* had smaller ovarian follicles than did nonallatectomized long-photophase females, and the same size as those of nonallatectomized short-photophase females. Further, topical application of synthetic JH resulted in follicle growth in short-photophase

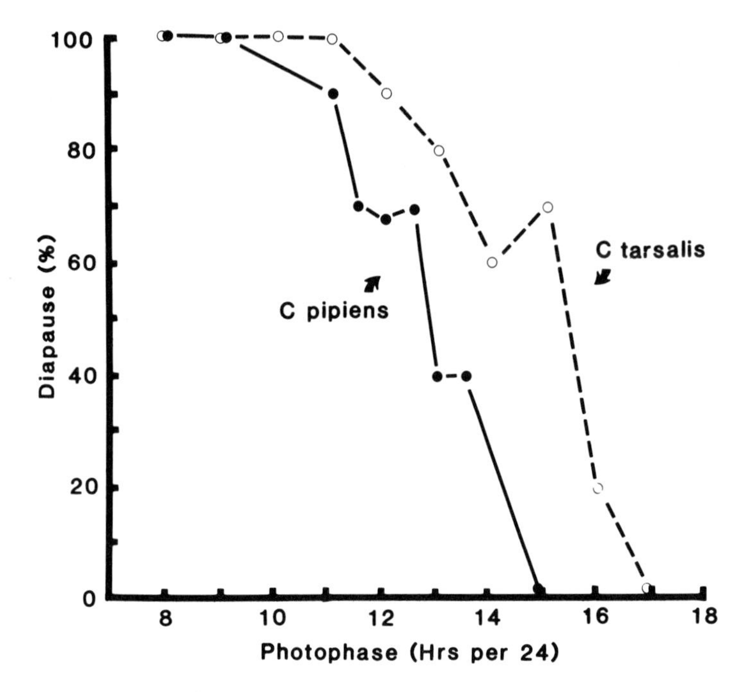

FIGURE 1.2. Percent diapause (follicle:germarium ratio 1.5:1 or less) in western Oregon strains of *Culex pipiens* and *Culex tarsalis* at 18°C and various photoperiods.

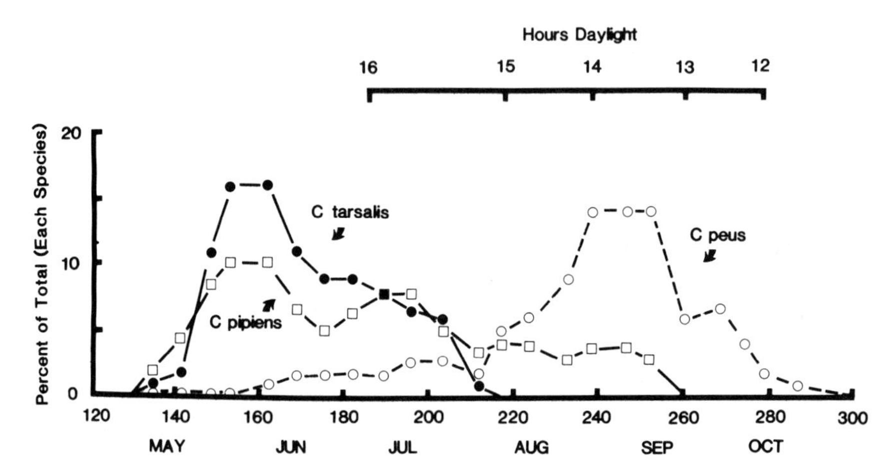

FIGURE 1.3. Results of season-long collections of mosquito larvae in a western Oregon log pond, 1981. Each symbol represents a value derived from a 3-week moving average of larvae collected. Julian dates are shown on abcissa. Hours of daylight, including civil twilight, are shown for this area (45.5°N latitude).

females. Whether JH directly controls blood feeding drive seems less certain. Case *et al.* (23) showed that topical application or feeding of a JH-mimic in diapausing female *Anopheles freeborni* Aitken resulted in resumption of blood-feeding. Meola and Petralia (64) showed that female *C. pipiens* and *C. quinquefasciatus* allatectomized within 1 hr of emergence failed to take blood meals, and that blood-feeding drive was restored after injection of JH-I. In *A. aegypti* (54) and *A. freeborni* (7) blood-feeding drive may be controlled by a factor secreted from developing ovaries. Also, in *A. aegypti,* females allatectomized at emergence do not have reduced blood-feeding drive (53). Meola and Petralia (64) ruled out a direct influence of an ovarian factor in *C. pipiens* and *C. quinquefasciatus,* however, by showing that there was no difference in blood-feeding behavior between females ovariectomized within 1 hr of emergence and nonovariectomized females.

Even if JH does directly control ovarian development and blood avidity in young (1- to 2-day-old) unfed females, there are probably quantitative differences involved. Although varying response to JH titers by different physiological mechanisms has not been demonstrated for mosquitoes, it has been shown to occur in roaches (9). If different levels of the same hormone control both blood-feeding drive and previtellogenic ovarian development, an important question is which level controls which function. In *C. pipiens,* Eldridge and Bailey (34) showed that when diapausing females were warmed at 25°C for 72–84 hr, 40% of those females that took blood meals had diapausing ovaries that did not subsequently develop. This would suggest that a higher level of JH is needed for ovarian development than for blood avidity.

Levels of JH have been estimated by various methods. Baker *et al.* (5) used an assay based on ovarian follicle response. Guilvard *et al.* (38) used a radioimmune assay to measure JH-3 levels in *Aedes detritus* (Haliday) and *A. caspius* (Pallas) (autogenous strains of both). In both species they found a pronounced peak of ecdysteroids at about 40 hr postemergence, followed by a peak of JH-3 at 48 hr. Unfortunately, no other biological factors were measured in parallel with the hormone titers (e.g., ovarian follicle growth). Studies are badly needed that employ such methods in diapausing and nondiapausing *Culex*. Such studies will firmly establish the role of hormone types and levels on physiological factors related to diapause.

Diapause Maintenance

Hibernacula

Hibernating *Culex* mosquitoes have been collected in a number of natural and man-made habitats. Natural sites have included talus slopes (40, 91), rock slides (90), rock crevices and caverns (44), caves (72, 116), brush piles (20), and wild animal burrows (12, 98). Man-made sites have included mines (11, 18, 25), concrete bridges (12), underground ammunition bunkers (3, 50), and basements and cellars (13, 52, 113). Although reports of crawl spaces under houses, sheds, and other above-ground sites have been reported to serve as hibernacula, such habitats do not usually provide satisfactory conditions for successful hibernation. In searches for overwintering mosquitoes between 1969 and 1976, co-workers and I frequently found *Culex* in above-ground structures in eastern Maryland and Virginia through December, but with a single exception, all such specimens were gone by January. The exception was a single *C. salinarius* female we collected in a livestock shelter in Accomack County, Virginia in January (36).

There have been a few detailed studies of the microenvironments present in hibernacula based on conditions recorded over an entire winter (52, 60, 65, 98). More numerous are reports based one or more single point measurements of temperature and humidity. Kuchlein and Ringelberg (55) discussed this subject in detail. Most hibernacula described have the following characteristics: very high relative humidity, little or no air movement, and cold, but not freezing, temperatures (52, 55, 62). An optimum hibernaculum temperature would appear to be ~5°C. There is little evidence to indicate that *Culex* mosquitoes can withstand subfreezing temperatures for extended periods and Lomax (60) found that average mortality per day increased as hibernacula temperatures approached 0°C in underground concrete bunkers. Mitchell (65) recorded temperatures as low as −3.9°C in mines containing hibernating *C. tarsalis,* but suggested that temperatures this cold and the accompanying low relative humidity (46–59%) were important mortality factors. On the other hand, if temperatures are too high, fat reserves would be prematurely expended (1). Even under the best of conditions overwinter mortality is high from predation, freezing, fat depletion, and desiccation (4, 50, 59, 108). Mitchell (65) reported posthibernation recovery of 8.7% of females of *C. tarsalis* marked and released in late fall in a Colorado mine. Bailey *et al.* (4) studied the survival of *C. pipiens* females in underground concrete bunkers in Maryland. They compared survival in six groups. They considered the first three groups to be in a state of diapause: (1) wild-collected diapausing females; (2) bloodfed, nongravid females laboratory-reared under diapause-inducing conditions; and (3) nonbloodfed females so reared. They considered the second three groups to be gonoactive (nondiapausing): (4) bloodfed, gravid females laboratory-reared under diapause-inducing conditions; (5) nonbloodfed females reared under nondiapause-inducing conditions; and (6) bloodfed, gravid females so reared. They found that the first three groups successfully overwintered, and survived significantly longer than did the last three groups, which did not survive the winter. These results demonstrate that bloodfed females can overwinter successfully if they are in diapause.

Physiological Changes during Diapause

During the course of the winter, fat reserves gradually become metabolized. Buxton (22) showed that the percentage of fat in diapausing *C. pipiens* dropped from 61% of dry body weight in October to 18% in March. He found that the amount of water present in diapausing females gradually decreased as well. Buffington and Zar (19), Schaefer and Washino (93), and Schaefer *et al.* (95) obtained similar results, but with specific information on lipid composition. Schaefer and Washino (93) found that total lipids during winter consisted mostly of triglycerides and free fatty acids. Schaefer and Washino (94) also collected samples of overwintering *C. apicalis* Adams females in California and found that individuals sampled during the winter contained 0.19–0.30 mg of total lipids.

During the winters of 1980–1983 I periodically collected diapausing females of *C. pipiens* and *C. tarsalis* from under concrete bridges and *C. pipiens* females from underground ammunition bunkers in Benton County, Oregon. Up to 20 females were removed at approximately 2-week intervals from September until April. The mosquitoes from each collection were divided into two equal groups. Individuals from one group were dissected for determination of ovarian follicle size, those from the other were extracted in petroleum ether in a Soxhlet apparatus to determine the amount of fat present. The results are shown in Figures 1.4 and 1.5. During 1980–1981, total lipids in *C. pipiens* females consisted of 60% of dry body weight in September (Fig. 1.4). By mid-April, when the last collections were made, fat consisted of 28% of dry body weight. During 1982–1983, the results were very similar. Values for *C. tarsalis* (Fig. 1.5) fluctuated more than those of *C. pipiens* in both 1981–1982 and 1982–1983. They were found to have lipids making up over 50% of their dry body weight at the time they first appeared under bridges. The percentage fell gradually over the winter until early February, when the last females were present. At this time fat made up 24% of dry body weight.

Little is known about the change of status of ovarian follicles over winter. Oda (72) examined follicles of *C. pipiens pallens* Coquillett collected in caves in the vicinity of Nagasaki, Japan at various times during the winter. He found that follicles increased in size gradually over the winter and that by March, all females sampled had "common-sized" (nondiapausing) follicles.

Diapause Termination

Very little is known about the factors that terminate diapause in *Culex* mosquitoes. Beck (8) discusses the term "diapause development" and suggests that expressions such as "breaking diapause" and "reactivation" are misleading because they suggest that the diapausing insect is in a com-

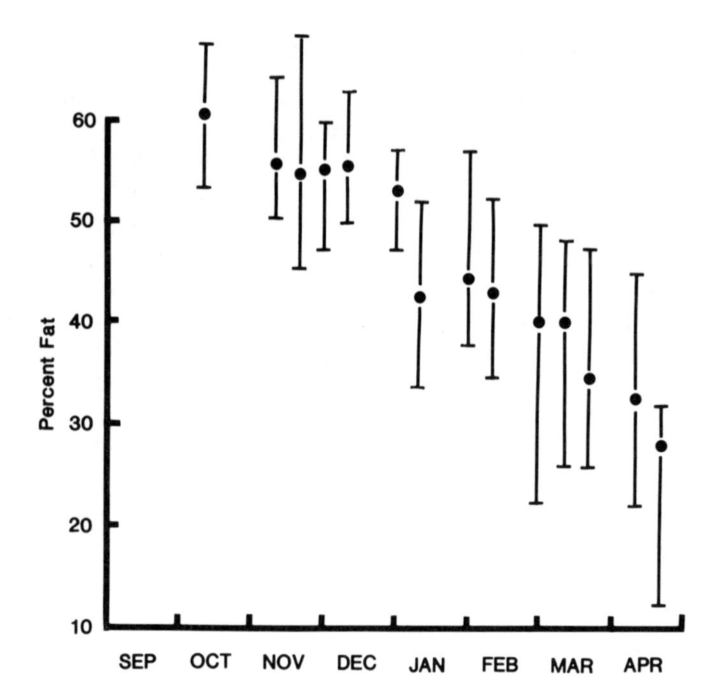

FIGURE 1.4. Percent fat in samples of *Culex pipiens* females collected under bridges and in an abandoned ammunition bunker, Benton County, Oregon, 1981–1982. Mean and range of values for samples of 10 females are shown.

pletely inactive state from which it must be roused. Diapause development, on the other hand, suggests a series of physiological events that must take place, after which the insect moves on to the nondiapause state. This concept deemphasizes the possible importance of temperature and photoperiod in ending diapause in mosquitoes. In reviewing the literature on diapause termination in mosquitoes, it is probably best to consider papers in two categories. Most of the papers apply (intentionally or not) to a concept of premature abortion of diapause, either as a control strategy (104) or as a potential natural fall season phenomenon (34, 66). A few papers, such as that of Bennington *et al.* (11), are based on studies of natural populations of mosquitoes emerging in spring.

In the former category, Eldridge (30) showed that *C. tritaeniorhynchus* would resume blood feeding several days after transferring diapausing females to a long-day photoperiod cycle. In *C. pipiens,* warm temperatures alone will terminate diapause (34). This provides an explanation for the apparent discrepancy between the results of Eldridge (32), on the one hand, and Oda and Wada (76) and Sanburg and Larsen (92), on the other. The former reported the lack of ovarian development in bloodfed female *C. pipiens* that had been subjected to short photophases. The latter investigators reported ovarian development under such conditions. Females

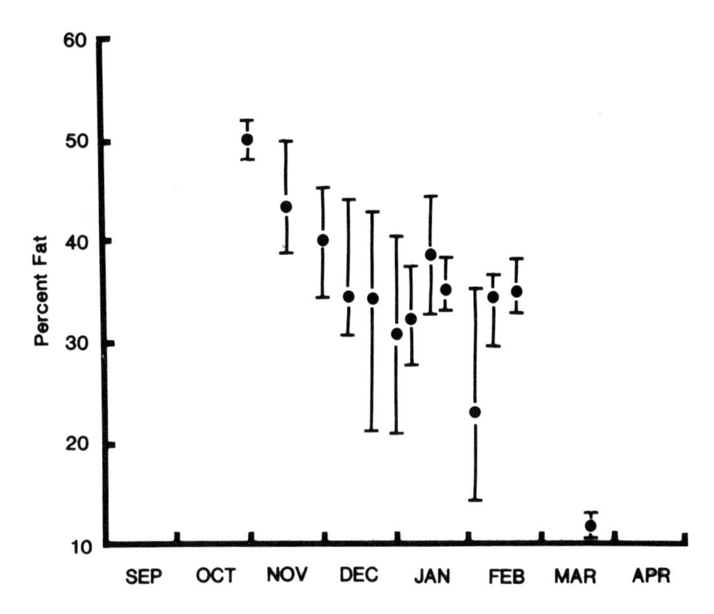

FIGURE 1.5. Percent fat in samples of *Culex tarsalis* females collected under bridges, Benton County, Oregon, 1981–1982. Mean and range of values for samples of 10 females are shown.

tested by Eldridge (32), however, were held at 15°C while blood was digested, while those tested by Oda and Wada (76) and by Sanburg and Larsen (92) were held at 21°C. Mitchell (66) found that failure of ovaries to develop in bloodfed female *C. tarsalis* depended upon incubation at short-photophase, cool temperature conditions. Thus it appears that higher temperatures promote JH synthesis, whereas short photophases inhibit it. Experiments designed to study JH-mediated events must examine both temperature and photoperiod conditions carefully, and over extended periods of time.

A few studies apply to populations of *Culex* mosquitoes that have gone through an entire winter in diapause. Bennington *et al.* (11) studied the relationship between soil temperatures and the initial appearance of *C. tarsalis* in spring for 4 years in Colorado. They found that emergence corresponded well with the time that soil surface temperatures became higher than subsurface temperatures (the soil temperature inversion). They concluded that temperature alone governed emergence from hibernacula. This was confirmed nicely by Shemanchuk (98). It does not follow from these observations, of course, that temperature alone is responsible for termination of diapause.

Oda (72) examined ovarian follicles of *C. pipiens pallens* females he collected overwintering in caves in Nagasaki, Japan. The presentation of his data, however, do not provide clues to possible factors terminating diapause. Earlier, I described the periodic winter collections of *C. tarsalis* and *C. pipiens* made under bridges in western Oregon. Dissections of collected females permitted the monitoring of ovarian diapause during the course of the winter. *Culex pipiens* females (Fig. 1.6) had diapause follicles until March. The growth curve was not linear, however, and was almost flat for most of the winter. The time at which the growth curve turns up in the spring agrees well with the critical photophase determined experimentally (12.71 hr.). In contrast, follicle size of *C. tarsalis* females (Fig. 1.7) increased linearly over the winter, with most females having nondiapause follicles by mid-February. This pattern is not consistent with the critical photophase (14.81 hr.) determined for this strain of *C. tarsalis*. The data shown are for the 1981–1982 winter. Since the patterns of follicle growth for both species were very similar during the three winters collections were made, I combined the data and subjected them to regression analysis. The data for *C. tarsalis* best fit a straight line with a formula $Y = 0.044 + 0.0002X$. For *C. pipiens,* on the other hand, the data best fit a third-order polynomial

$$Y = 0.047 + 2.99X - 0.03X^2 + 0.000094X^3$$

These studies suggest that follicles of the western Oregon strain of *C. tarsalis* grow gradually over the winter irrespective of photoperiod, and that by late winter, ovarian diapause is completed. Ovaries of *C. pipiens*

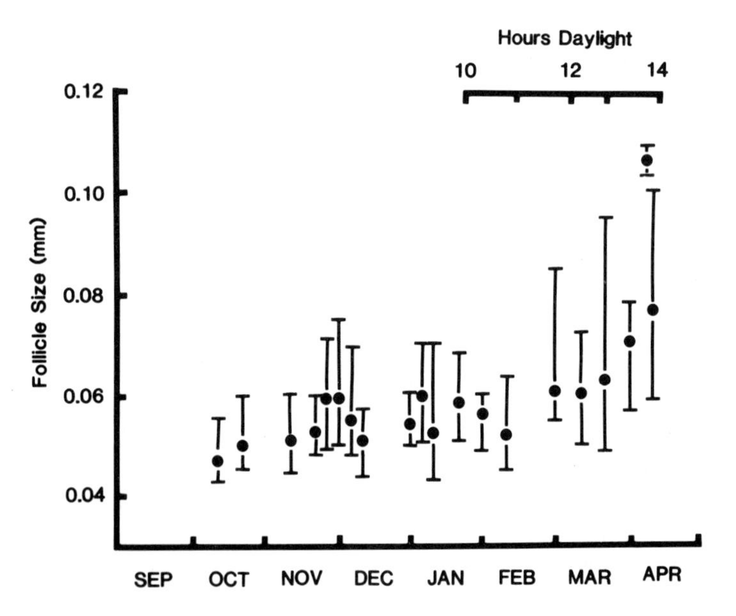

FIGURE 1.6. Ovarian follicle length in samples of *Culex pipiens* females collected under bridges and in an abandoned ammunition bunker, Benton County, Oregon, 1981–1982. Symbols represent mean and range of values for samples of 10 females in most instances; some April values are based on smaller samples. Hours of daylight, including civil twilight, are shown for this area (45.5°N latitude).

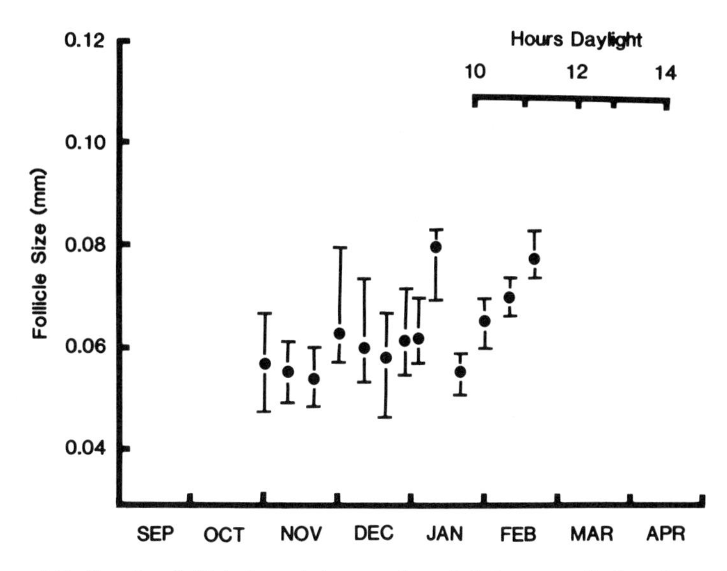

FIGURE 1.7. Ovarian follicle length in samples of *Culex tarsalis* females collected under bridges, Benton County, Oregon, 1981–1982. Symbols represent mean and range of values for samples of 10 females in most instances; some February values are based on smaller samples. Hours of daylight, including civil twilight, are shown for this area (45.5°N latitude).

from this same area seem to remain in diapause until spring, when diapause is terminated perhaps by increased day length. Oda and Kuhlow (74) suggest that day length terminates diapause in *C. pipiens* in northern Germany. Spielman and Wong (106) found that most female *C. pipiens* collected in the fall in the area of Boston, Massachusetts had nondiapause follicles by mid-October. These may have been representatives of an autogenous strain, however.

Autogeny

The fact that females of some strains or populations of mosquitoes can produce viable eggs without a previous blood meal has been known for many years (103). First reported for *C. pipiens,* it is now known to occur in many mosquito species in various genera (26). Genetics of autogeny in *Culex* mosquitoes has been studied in detail only in *C. pipiens*. In this species it is controlled by several gene loci (103). In other *Culex* species it is also under genetic control, but its expression also can be modified in populations by environmental factors (29, 41, 79, 103). Autogenous strains of *Culex pipiens* do not enter diapause. Spielman and Wong (105) found that autogenous strains of this species did not undergo ovarian diapause under conditions that produced 97% diapause in an anautogenous population. Interbreeding among anautogenous and autogenous strains of this species occurs at a very low level (102), so that biological characteristics in the species are fixed to a fairly high degree (stenogamy, autogeny, lack of diapause). Barr (6) reviewed autogeny in the *C. pipiens* complex, including the question of genetic control of autogeny and other biological characteristics of autogenous strains.

Such characterization of autogenous and anautogenous strains is much more difficult in other species, such as *C. tarsalis*. In this species, autogeny is expressed to varying degrees according to the time of year (68), the photoperiod (41), temperature, and various factors affecting larval nutrition. For such species, the concept that autogenous strains do not enter diapause would seem to need closer examination. The expression of autogeny and that of ovarian diapause are mutually exclusive physiological events, and one may assume that an individual in which autogeny is expressed will not enter diapause. Harwood (41) reported that exposure of larval and pupal stages to short photophases suppressed expression of autogeny in adult female *C. tarsalis*, and Reisen (83) achieved similar results.

In order to shed some light on the question of diapause potential in autogenous strains of *C. tarsalis*, and also to examine ovarian diapause in different geographical strains of the species, I studied three colonized strains from California. Dr. William Reisen of the University of California Bakersfield Field Station provided material from Kern County, California (Breck-80 strain) and from Yolo County, California (Knight's Landing strain). Autogeny occurred in about 50% of the individuals in each strain. Dr. Robert K. Washino of the University of California at Davis provided a laboratory strain that had been colonized many years previously from material collected in the Davis area. Autogeny in this strain was about 90% (it had been selected for autogeny for many years). All three strains were subjected to four combinations of temperature and photoperiod: 15°C, L:D 8:16; 15°C, L:D 16:8; 25°C, L:D 8:16; and 25°C, L:D 16:8. In all cases pupae were separated randomly and placed in incubators programmed for these conditions, and emerged adults were held for 8 days. In the case of the Davis strain, ten females were removed each day from each experimental group and dissected for determination of ovarian follicle size. Autogenous females could be recognized after 3–6 days of exposure, depending upon temperature, because they had follicles that had grown beyond Stage IIb of Kawai (51) (about 0.12 mm in this species). Those that had not grown beyond that stage were considered anautogenous and were recorded separately. Ovaries of females that had deposited a clutch of eggs autogenously could likewise be detected by the presence of a follicular relic on the pedicel, and the length of the secondary follicles was compared for such individuals. The results of dissections of females of the Davis strain are shown in Table 1.2. Based on the determination of autogeny, as described above, the percentage of autogenous females at each treatment condition is shown in Table 1.3.

TABLE 1.2. Ovarian follicle length (mm) of ovaries of female *Culex tarsalis* (Davis strain) on each of 8 days postemergence.[a]

Temperature (°C) and photoperiod	Follicle length at given day postemergence							
	1	2	3	4	5	6	7	8
All follicles								
15°C, 8:16	0.04	0.06	0.08	0.11	0.15	0.12	0.17	0.20
15°C, 16:8	0.04	0.05	0.08	0.10	0.15	0.24	0.23	0.29
25°C, 8:16	0.06	0.12	0.30	0.51	0.53	0.55	0.54	0.54
25°C, 16:8	0.05	0.13	0.15	0.28	0.52	0.44	0.51	0.48
Anautogenous follicles only								
15°C, 8:16	—	—	—	—	—	0.08 (6)	0.10 (4)	0.10 (3)
15°C, 16:8	—	—	—	—	—	0.14 (3)	0.12 (5)	0.13 (5)
25°C, 8:16	—	—	—	0.11 (1)	0.10 (1)	—	—	0.13 (1)
25°C, 16:8	—	—	—	—	—	0.13 (1)	0.13 (1)	0.13 (2)
Secondary follicles only								
25°C, 8:16	—	—	—	0.05 (4)	0.06 (6)	0.07 (6)	—	0.06 (7)
25°C, 16:8	—	—	—	—	—	—	0.06 (9)	0.07 (5)

[a]Pupae and adults maintained under conditions indicated. Refer to text for methods used to differentiate among autogenous, anautogenous, and secondary follicles. Values are means based on 10 individual females unless indicated otherwise, in which case the number of females is shown in parentheses under the mean value.

Material from the Breck-80 and Knight's Landing strains were handled somewhat differently. After exposure of pupae and adults to the four photoperiod/temperature combinations, each group of adults was divided into two subgroups at day 8 of adult life. One subgroup from each treatment was dissected to determine degree of ovarian development; the other was exposed to a restrained Japanese quail in a 1-ft^3 cage overnight. Females that fed on the quail were held an additional 8–10 days at the same treatment conditions, at which time they were dissected. The results of these dissections are shown in Table 1.4 for the Knight's Landing strain and Table 1.5 for the Breck-80 strain. For both the Knight's Landing strain

TABLE 1.3. Percent autogeny in female *Culex tarsalis* (Davis strain) maintained under four conditions of temperature and photoperiod.

Temperature (°C) and photoperiod	Number determinable	Number autogenous	Percent
15°C, 8:16	30	17	56.7
15°C, 16:8	30	15	50.0
25°C, 8:16	60	57	95.0
25°C, 16:8	60	56	93.3

TABLE 1.4. Ovarian follicle size in bloodfed and nonbloodfed *Culex tarsalis* (Knight's Landing strain) maintained for 8 days under four combinations of temperature and photoperiod.

Temperature (°C) and photoperiod	Percent autogenous	Percent bloodfed	Follicle length (mm)
Nonbloodfed			
15°C, 8:16	ND[a]	—	0.12
15°C, 16:8	ND[a]	—	0.10
25°C, 8:16	60	—	0.10
25°C, 16:8	20	—	0.10
Bloodfed			
15°C, 8:16	—	50	0.54
15°C, 16:8	—	100	0.53
25°C, 8:16	—	60	0.51
25°C, 16:8	—	100	0.50

[a]Not done.

and the Breck-80 strain, there was no evidence of ovarian diapause among the nonbloodfed groups. Also, in both strains, the proportion of females taking blood was significantly higher in the long-photoperiod groups at both temperatures. In all cases for both strains, ovarian development proceeded to maturity. These data raise interesting questions. First of all, even among individuals that had anautogenous development, there was no significant difference in ovarian follicle development among temperature and photoperiod treatment conditions for any of the three strains. Does this mean that these individuals had an unexpressed genetic capacity for autogeny and that such individuals also do not have the capacity for diapause? It is also interesting to note that in females developing eggs autogenously, the follicles are at the pre-resting stage (No2) at the time of

TABLE 1.5. Ovarian follicle size in bloodfed and nonbloodfed *Culex tarsalis* (Breck-80 strain) maintained for 8 days under four combinations of temperature and photoperiod.

Temperature (°C) and photoperiod	Percent autogenous	Percent bloodfed	Follicle length (mm)
Nonbloodfed			
15°C, 8:16	ND[a]	—	0.09
15°C, 16:8	ND[a]	—	0.09
25°C, 8:16	50	—	0.10
25°C, 16:8	40	—	0.10
Bloodfed			
15°C, 8:16	—	40	0.50
15°C, 16:8	—	100	0.60
25°C, 8:16	—	89	0.70
25°C, 16:8	—	82	0.59

[a]Not done.

emergence; in this regard they do not differ from anautogenous strains. In the Davis strain, the percentage of females expressing autogeny varied significantly between temperature treatments, but not between photoperiod treatments. In this they do not agree with the results of Harwood (41) using a Washington state strain, nor of Reisen (83) studying a strain from Kern County, California. The Davis strain had been under colonization for many years, however, and may have lost its ability to enter diapause. Reisen (85) suggests that colonization and insectary culture may select against ability to enter diapause.

Overwinter Survival of Arboviruses

This subject has been covered several times previously (33, 81), with reviews of possible overwintering mechanisms other than persistence of the pathogens in their arthropod vectors. I will restrict this discussion to the question of the perpetuation of togaviruses in *Culex* vectors. Two related questions remain unanswered in this regard. First, do these pathogens usually persist over winter in temperate regions in infected mosquito vectors, and second, if they do, how do the mosquitoes become infected? The most obvious direct evidence relating to the first question is the isolation of virus from diapausing female mosquitoes during the winter. In spite of numerous attempts to do so, there have been relatively few isolations (3, 14, 43, 58, 82, 111). In light of current knowledge of the biology of arboviruses, these past failures are not surprising. Past attempts to isolate virus from overwintering mosquitoes were made on the assumption that conventional primary virus isolation techniques (mouse or tissue culture inoculation) used to detect virus during summer periods when virus is active in vertebrate and invertebrate cycles would also work for winter isolations. This assumption may have been wrong. The isolations of SLE by Bailey *et al.* (3) were obtained only from female mosquitoes that had been warmed and then given a noninfective blood meal after their collection. Based on studies of alphaviruses in tissue culture, Schlesinger *et al.* (97) postulate that an arbovirus, after prolonged replication in mosquitoes, will become progressively more temperature sensitive and avirulent. Ability to detect and isolate viruses from infected mosquitoes is dependent upon the virulence of the virus for the biological system used. Such virulence is now thought to be a highly complex phenomenon, dependent upon such factors as relative amounts of temperature-sensitive mutant particles present, production of defective-interfering genomes and virions, specific immune responses, and perhaps interferon production (96, 97, 107). Thus, it is possible that infections in diapausing mosquitoes may differ in some of these regards from gonoactive mosquitoes present during the active transmission season. Because of these factors, further attempts should be made to isolate viruses from diapausing female mosquitoes, but by varying temperatures, providing uninfected blood meals, and perhaps performing other experimental manipulations first. More sensitive methods of virus detection, such as the use of radioimmune assays, ELISA, or cDNA probes, need to be developed for such studies as well.

The second question remains open, that is, how do the diapausing females become infected? It seems almost certain now that California serotype viruses are maintained in their invertebrate vectors through transovarial transmission (33). Although transovarial transmission of these bunyaviruses has been demonstrated in aedine mosquitoes that overwinter in the egg stage, proof of that togaviruses are transovarially transmitted in *Culex* vectors has been more elusive, especially in the case of alphaviruses. Low rates of transovarial transmissin of SLE have been reported in *C. quinquefasciatus,* however, (37), and of JE in *C. tritaeniorhynchus* (87). Recently Hardy *et al.* (39) showed that *C. tarsalis* can transmit virus to their progeny when reared at 18°C, but not when reared at 27°C. Although the rates of transovarial transmission in all of these experiments would seem to be too low to represent a frequently occurring natural phenomenon, it is possible that transmission efficiency in nature is much higher than what has been demonstrated in experimental situations. It is also possible that in nature stabilized infections of germ cells of female mosquitoes exist, as shown for San Angelo virus in *Aedes epactius* Dyar and Knab (110). This mechanism of infection of diapausing female mosquitoes thus cannot be discounted. The essential proof that is missing, however, is the isolation of virus from *Culex* larvae or males from nature, preferably in the late summer. More sensitive methods of virus detection may permit demonstration of higher rates of transovarial transmission and isolation of virus from larvae.

The other means by which hibernating *Culex* mosquitoes might become infected is by biting an infected vertebrate host. This has generally been considered unlikely because prehibernating females seem not to take a blood meal. Diapause in *Culex* mosquitoes is marked by ovarian diapause and reduced blood avidity. For a prehibernating female to take an infectious blood meal and successfully hibernate, the following events would have to take place: (1) ovarian diapause; (2) an infectious blood meal, in spite of reduced avidity; (3) lack of ovarian development; and (4) hypertrophy of the fat body. These conditions can be produced experimentally by warming diapausing females for 48–84 hr (34), but Mitchell (66, 67) has shown that diapausing female *C. tarsalis* and *C. pipiens* will take blood meals if confined to small (0.24 liter), but not large (3.8 liter) feeding containers. His interpretation of these results is that feeding tests that take place in very small containers circumvent the need for questing behavior, and that large-container tests more nearly simulate natural conditions. It is now accepted that diapausing female *Culex* ovaries will remain undeveloped should a blood meal be taken, and nulliparity in overwintering females in nature cannot be used as evidence of lack of a prior blood meal. Because all laboratory tests are at best approximations of natural conditions, evidence of blood feeding (or lack of blood feeding) in prehibernating females in nature is needed to settle this question. This will probably await the development of an immunological test of some kind that can detect blood meal products after complete digestion of the blood meal. However, Bailey *et al.* (4) report collecting a bloodfed *C. pipiens* in a wooden shelter in Maryland on November 1. The female was taken to a laboratory and maintained at 15°C, L:D 9:15. The ovaries did not develop. This is one of the few demonstrations of failure of ovaries to develop fully in a blood-engorged *Culex* mosquito in nature, and it points out that even if it is a rare occurrence, occasional blood feeding by prehibernating female *C. pipiens* does occur.

At present, attempts to develop a model for overwinter survival in the vector of *Culex*-transmitted arboviruses must deal with three events that seem to occur only rarely: isolation of virus from overwintering female mosquitoes, transovarial transmission of virus, and blood feeding in prehibernating females. New technology may facilitate attempts to explore the natural occurrence of all three of these factors, but conventional approaches continue to be useful. For example, we need studies of the behavior of prehibernating females in nature, especially when they are actively seeking carbohydrate sources for hypertrophic fat. At the same time, new methods must be applied to understand the physiological nature of diapause in *Culex* mosquitoes. Since the entire question of overwinter survival of virus is a matter of probability analysis, a mathematical systems approach seems called for. So far, however, not enough reliable data are available to form the basis for the analysis.

Summary and Conclusions

In spite of a very large amount of research conducted over the last 50–70 years on diapause phenomena in *Culex* mosquitoes and of the potential role of mosquito vectors as overwinter reservoirs of virus, basic questions remain. Although the ecological facets of diapause are reasonably well understood for a few species of *Culex,* such aspects as endocrine control of diapause are not well known at all. More comparative ecological research is needed, as well as more physiological research. This research needs to be conducted in both laboratory and field settings. These endeavors can incorporate new techniques of immunology, virology, and entomology. Thus, we are left to speculate as to what happens in nature based on rare and sometimes conflicting demonstrations in the laboratory of key events.

Acknowledgments. Thanks are due to many individuals who participated in laboratory and field studies reported here. Anne Tempte, Scott Ritchie, Diane Henneberger, George Hoffman, David Walter, Intawat Burikam, and Katherine Macdonald all made valuable contributions. I also thank Andrew Spielman, William Reisen, and René Feyereisen for their critical review of the manuscript and for their discussion of the topics covered. Studies reported here were supported by National Institutes of Health Grant AI 16346. This is Oregon Agricultural Experiment Station Technical Paper 7531.

References

1. Anderson, A.W., and Harwood, R.F. 1966, Cold tolerance in adult female *Culex tarsalis* (Coquillett), *Mosq. News* **26:**1–7.
2. Anderson, W.A., and Spielman, A., 1971, Permeability of the ovarian follicle of *Aedes aegypti* mosquitoes, *J. Cell Biol.* **50:**201–221.
3. Bailey, C.L., Eldridge, B.F., Hayes, D.E., Watts, D.M., Tammariello, R.F., and Dalrymple, J.M., 1978, Isolation of St. Louis encephalitis virus from overwintering *Culex pipiens* mosquitoes, *Science* **199:**1346–1349.
4. Bailey, C.L., Faran, M.E., Gargan, T.P., and Hayes, D.E., 1982, Winter survival of blood-fed and non blood-fed *Culex pipiens* L., *Am. J. Trop. Med. Hyg.* **31:**1054–1061.
5. Baker, F.C., Hagedorn, H.H., Schooley, D.A., and Wheelock, G., 1983, Mosquito juvenile hormone: Identification and bioassay activity, *J. Insect Physiol.* **29:**465–470.
6. Barr, A.R., 1982, The *Culex pipiens* complex, in: W.W.M. Steiner, W.J. Tabachnick, K.S. Rai, and S. Narang (eds.), *Recent Developments in the Genetics of Insect Disease Vectors,* Stipe, Champaign, Illinois, pp. 551–572.
7. Beach, R. 1979. Mosquitoes: Biting behavior inhibited by ecdysone, *Science* **205:**829–831.
8. Beck, S.D., 1980, *Insect Photoperiodism,* 2nd ed., Academic Press, New York. 387 pp.
9. Bell, W.J., and Barth, R.H., Jr., 1970. Quantitative effects of juvenile hormone on reproduction in the cockroach *Byrsotria fumigata, J. Insect Physiol.* **16:**2302–2313.
10. Bellamy, R.E., and Reeves, W.C., 1963, The winter biology of *Culex tarsalis* (Diptera: Culicidae) in Kern County, California, *Ann. Entomol. Soc. Am.* **56:**314–323.
11. Bennington, E.E., Blackmore, J.S., and Sooter, C.A., 1958a. Soil temperature and emergence of *Culex tarsalis* from hibernation, *Mosq. News* **18:**297–298.
12. Bennington, E.E., Sooter, C.A., and Baer, H., 1958b. The diapause in adult female *Culex tarsalis* Coquillett, *Mosq. News* **18:**299–304.
13. Berg, M., and Lang, S., 1948, Observations of hibernating mosquitoes in Massachusetts, *Mosq. News* **8:**70–71.
14. Blackmore, J.S., and Winn, J.F., 1956, A winter isolation of western equine encephalitis virus from hibernating *Culex tarsalis* Coq., *Soc. Exp. Biol. Med. Proc.* **91:**146–148.
15. Boissezon, P. de., 1929, Remarques sur le conditions de la reproduction chez *Culex pipiens* L. pendant la periode hivernale, *Bull. Soc. Pathol. Exot.* **22:**549–552.

16. Boissezon, P. de., 1930, Influence de la temperature sur la biologie des culicides, *Bull. Soc. Zool. Fr.* **55:**155–261.
17. Boissezon, P. de., 1934, Nouvelles experiences sur la biologie de *Culex pipiens* L., *Ann. Parasitol. Hum. Comp.* **12:**182–192.
18. Buffington, J.D., 1972, Hibernaculum choice in *Culex pipiens, J. Med. Entomol.* **9:**128–132.
19. Buffington, J.D., and Zar, J.H., 1968, Changes in fatty acid composition of *Culex pipiens pipiens* during hibernation, *Ann. Entomol. Soc. Am.* **61:**774–775.
20. Bullock, H.R., Murdoch, W.P., Fowler, H.W., and Brazzel, H.R., 1959, Notes on the overwintering of *Culex tritaeniorhynchus* Giles in Japan, *Mosq. News* **19:**184–188.
21. Burdick, D.J., and Kardos, E.H., 1963, The age structure of fall, winter, and spring populations of *Culex tarsalis* in Kern County, California, *Ann. Entomol. Soc. Am.* **56:**527–535.
22. Buxton, P.A., 1935, Changes in the composition of adult *Culex pipiens* during hibernation, *Parasitology* **27:**263–265.
23. Case, T.J., Washino, R.K., and Dunn, R.L., 1977, Diapause termination in *Anopheles freeborni* with juvenile hormone mimics, *Entomol. Exp. Appl.* **21:**155–162.
24. Chapman, H.C., 1959, Overwintering larval populations of *Culex erythrothorax* in Nevada, *Mosq. News* **19:**244–246.
25. Chapman, H.C., 1961, Abandoned mines as overwintering sites for mosquitoes, especially *Culex tarsalis* Coq. in Nevada, *Mosq. News* **21:**234–237.
26. Clements, A.N., 1963, *The Physiology of Mosquitoes,* Macmillan, New York. 393 pp.
27. Clements, A.N., and Boocock, M.R., 1984, Ovarian development in mosquitoes: Stages of growth and arrest, and follicular resorption, *Physiol. Entomol.* **9:**1–8.
28. Danilevskii, A.S., and Glinyananya, E.I., 1958, The dependence of the gonotrophic cycle and imaginal diapause of blood sucking mosquitoes on variation in day-length, *Uch. Zap. Leningr. Gos. Univ. 240, Ser. Biol. Nauk* **46:**34–51.
29. Eberle, M.W., and Reisen, W.K., 1986, Studies on autogeny in *Culex tarsalis:*1a. Selection and genetic experiments, *J. Am. Mosq. Control Assoc.* **2:**38–43.
30. Eldridge, B.F., 1963, The influence of daily photoperiod on bloodfeeding activity of *Culex tritaeniorhynchus* Giles, *Am. J. Hyg.* **77:**49–53.
31. Eldridge, B.F., 1966, Environmental control of ovarian development in mosquitoes of the *Culex pipiens* complex, *Science* **151:**826–828.
32. Eldridge, B.F., 1968, The effect of temperature and photoperiod on bloodfeeding and ovarian development in mosquitoes of the *Culex pipiens* complex, *Am. J. Trop. Med. Hyg.* **17:**133–140.
33. Eldridge, B.F., 1981, Vector maintenance of pathogens in adverse environments (with special reference to mosquito maintenance of arboviruses), in: J.J. McKelvey, Jr., B.F. Eldridge, and K. Maramorosch (eds.), *Vectors of Disease Agents,* Praeger, New York, pp. 143–157.
34. Eldridge, B.F., and Bailey, C.L., 1979, Experimental hibernation studies in *Culex pipiens* (Diptera: Culicidae): Reactivation of ovarian development and blood-feeding in prehibernating females, *J. Med. Entomol.* **15:**462–467.

35. Eldridge, B.F., Bailey, C.L., and Johnson, M.D., 1972, A preliminary study of the seasonal geographic distribution and overwintering of *Culex restuans* Theobald and *Culex salinarius* Coquillett, *J. Med. Entomol.* **9**:133–238.
36. Eldridge, B.F., Johnson, M.D., and Bailey, C.L., 1976, Comparative studies of two North American mosquito species, *Culex restuans* and *Culex salinarius:* Response to temperature and photoperiod in the laboratory, *Mosq. News* **36**:506–513.
37. Francy, D.B., Rush, W.A., Montoya, M., Inglish, D.S., and Bolin, R.A., 1981, Transovarial transmission of St. Louis encephalitis virus by *Culex pipiens* mosquitoes, *Am. J. Trop. Med. Hyg.* **30**:699–705.
38. Guilvard, E., Reggie, M. de, and Rioux, J.-A., 1984, Changes in ecdysteroid and juvenile hormone titers correlated to the initiation of vitellogenesis in two *Aedes* species (Diptera: Culicidae), *Gen. Comp. Endocrinol.* **53**:218–223.
39. Hardy, J.L., Rosen, L., Reeves, W.C., Scrivani, R.P., and Presser, S.B., 1984, Experimental transovarial transmission of St. Louis encephalitis virus by *Culex* and *Aedes* mosquitoes, *Am. J. Trop. Med. Hyg.* **33**:166–175.
40. Harwood, R.F., 1962, Trapping overwintering adults of the mosquito *Culex tarsalis* and *Anopheles freeborni, Mosq. News* **22**:26–31.
41. Harwood, R.F., 1966, The relationship between photoperiod and autogeny in *Culex tarsalis, Entomol. Exp. Appl.* **9**:327–331.
42. Harwood, R.F., and Halfhill, J.E., 1964, The effect of photoperiod on fat body and ovarian development of *Culex tarsalis, Ann. Entomol. Soc. Am.* **57**:596–600.
43. Hayashi, K., Mifune, K., Shichijo, A., Suzuki, H., Matsuo, S., Makino, Y., Akashi, M., Wada, Y., Oda, T., Mogi, M., and Mori, A., 1975, Ecology of Japanese encephalitis virus in Japan. III. The results of investigation in Amami Island, southern part of Japan, from 1973 to 1975, *Trop. Med (Nagasaki)* **17**:129–142. [Cited in 82.]
44. Hayles, L.B., Weegnar, H.H., Iversen, J.O., and McLintock, J., 1979, Overwintering sites of adult mosquitoes in Saskatchewan, *Mosq. News* **39**:117–120.
45. Hecht, O., 1932, Experimentelle Beitrage zur Biologie der Stechmucken, *Z. Angew. Entomol.* **19**:578–607.
46. Hecht, O., 1933, Die Blutnahrung, die Erzeugung der Eier und die Uberwinterung der Stechmuckenweibehen, *Arch. Sch. Trop. Pathol. Ther. Exot. Krankh.* **37**(B 3):1–87.
47. Holzapfel, C.M., and Bradshaw, W.E., 1981, Geography of larval dormancy in the tree-hole mosquito, *Aedes triseriatus* (Say), *Can. J. Zool.* **59**:1014–1021.
48. Hudson, J.E., 1978, Overwintering sites and ovarian development of some mosquitoes in central Alberta, *Mosq. News* **38**:570–579.
49. Jordan, R.G., and Bradshaw, W.E., 1978, Geographic variation in the photoperiodic response of the western tree-hole mosquito, *Aedes sierrensis, Ann. Entomol. Soc. Am.* **71**:487–490.
50. Jumars, P.A., Murphey, F.J., and Lake, R.W., 1969, Can blood-fed *Culex pipiens* L. overwinter? *Proc. N.J. Mosq. Exterm. Assoc.* **56**:219–225.
51. Kawai, S., 1969, Studies on the follicular development and feeding activity of the females of *Culex tritaeniorhynchus* with special reference to those of autumn, *Trop. Med. (Nagasaki)* **11**:145–169.

52. Keener, G.G., Jr., 1952, Observations on overwintering of *Culex tarsalis* Coquillett (Diptera, Culicidae) in western Nebraska, *Mosq. News* **12:**205–209.
53. Klowden, M.J., 1983, The physiological control of mosquito host-seeking behavior, in K.H. Harris (ed.), *Current Topics in Vector Research,* Praeger, New York, Vol. 1, pp. 93–116.
54. Klowden, M.J., and Lea, A.O., 1979, Abdominal distention terminates subsequent host-seeking behaviour of *Aedes aegypti* following a blood meal, *J. Insect Physiol.* **25:**583–585.
55. Kuchlein, J.H., and Ringelberg, J., 1964, Further investigations on the distribution of hibernating *Culex pipiens pipiens* L. (Diptera, Culicidae) in artificial marlcaves in South Limburg (Netherlands), *Entomol. Exp. Appl.* **7:**25–46.
56. LaFace, L., 1926, Richerche sulla biologica del *Culex pipiens.* L'alimentazione e l'ivernamento, *Riv. Malariol.* **5:**132–156.
57. Lea, A.O., Briegel, H., and Lea, H.M., 1978. Arrest, resorption, or maturation of oocytes in *Aedes aegypti:* Dependence on the quantity of blood and the interval between blood meals, *Physiol. Entomol.* **3:**319–316.
58. Lee, H.W., 1971, Study on overwintering mechanisms of Japanese encephalitis virus in Korea, *J. Korean Med. Assoc.* **14:**871.
59. Lomax, J.L., 1967, Fall mosquito breeding and hibernation in an area of the Delaware River survey, *Proc. N.J. Mosq. Exterm. Assoc.* **54:**170–178.
60. Lomax, J.L., 1968, A study of mosquito mortality relative to temperature and relative humidity in an overwintering site, *Proc. N.J. Mosq. Exterm. Assoc.* **55:**81–85.
61. Madder, D.J., Surgeoner, G.A., and Helson, B.V., 1981, Induction of diapause in *Culex pipiens* and *C. restuans* (Diptera: Culcidae) in southern Ontario, *Can. Entomol.* **115:**877–883.
62. Mail, A.G., and McHugh, R.A., 1961, Relation of temperature and humidity to winter survival of *Culex pipiens* and *Culex tarsalis, Mosq. News* **21:**252–254.
63. Mansingh, D., 1971, Physiological classification of dormancies in insects, *Can. Entomol.* **103:**983–1009.
64. Meola, R.W., and Petralia, R.S., 1980, Juvenile hormone induction of biting behavior in *Culex* mosquitoes, *Science* **209:**1548–1550.
65. Mitchell, C.J., 1979, Winter survival of *Culex tarsalis* (Diptera: Culicidae) hibernating in mine tunnels in Boulder County, Colorado, USA, *J. Med. Entomol.* **16:**482–487.
66. Mitchell, C.J., 1981, Diapause termination, gonoactivity, and differentiation of host-seeking behavior from blood-feeding behavior in hibernating *Culex tarsalis* (Diptera: Culicidae), *J. Med. Entomol.* **5:**386–394.
67. Mitchell, C.J., 1983, Differentiation of host-seeking behavior from blood-feeding behavior in overwintering *Culex pipiens* (Diptera: Culicidae) and observations on gonotrophic dissociation, *J. Med. Entomol.* **20:**157–163.
68. Moore, C.G., 1963, Seasonal variation in autogeny in *Culex tarsalis* Coq. in northern California, *Mosq. News* **23:**238–241.
69. Nayar, J. K., 1982, Bionomics and physiology of *Culex nigripalpus* (Diptera: Culicidae) of Florida: An important vector of diseases, Florida Agricultural Experiment Station Bulletin No. 827. 73 pp.

70. Nelson, M.J., 1971, Mosquito studies (Diptera: Cullicidae) XXVI. Winter biology of *Culex tarsalis* in Imperial Valley of California, *Contrib. Am. Entomol. Inst.* **7:**1–56.
71. Nieschulz, O., and Bos, A., 1931, Einige versuche mit uberwinterden Exemplaren von *Culex pipiens, Zentralbl. Bakteriol.* **84:**364–368.
72. Oda, T., 1968, Studies on the follicular development and overwintering of the house mosquito, *Culex pipiens pallens* in Nagaski Area, *Trop. Med. (Nagasaki)* **10:**195–216.
73. Oda, T., 1971, On the effect of the photoperiod and temperature on the feeding activity and follicular development of *Culex pipiens pallens* females, *Trop. Med. (Nagasaki)* **13:**200–204.
74. Oda, T., and Kuhlow, F., 1974, Seasonal changes in gonoactivity of *Culex pipiens pipiens* in northern Germany and its response to day length and temperature, *Tropenmed. Parasitol.* **25:**175–186.
75. Oda, T., and Kuhlow, F., 1976, Gonotrophische Dissoziation bei *Culex pipiens* L. *Tropenmed. Parasitol.* **27:**101–105.
76. Oda, T., and Wada, Y., 1972, On the development of follicles after bloodfeeding in *Culex pipiens pallens* females where were reared under various enviromental conditions, *Trop. Med. (Nagasaki)* **14:**65–70.
77. Oda, T., and Wada, Y., 1973, On the gonotrophic dissociation in *Culex tritaeniorhynchus summarosus* females under various conditions, *Trop. Med. (Nagasaki)* **15:**189–195.
78. Oda, T. Wada, Y., and Mori, A., 1978, Follicular degeneration in unfed nulliparous females of *Culex tritaeniorhynchus, Trop. Med. (Nagasaki)* **20:**113–122.
79. O'Meara, G.F., 1979, Variable expression of autogeny in three mosquito species, *Int. J. Invertbr. Reprod.* **1:**253–261.
80. Peus, F., 1930, Zue Biologie der Hausmucke, *Culex pipiens* L., wahrend der Wintermonate, *Z. Desinfekt.* **22:**410–414.
81. Reeves, W.C., 1974, Overwintering of arboviruses, *Prog. Med. Virol.* **17:**193–220.
82. Reeves, W.C., Bellamy, R.E., and Scrivani, R.P., 1958, Relationships of mosquito vectors to winter survival of encephalitis vectors. I. Under natural conditions, *Am. J. Hyg.* **67:**78–79.
83. Reisen, W.K., 1986, Studies on autogeny in *Culex tarsalis:* 2. Simulated diapause induction and termination in genetically autogenous females, *J. Am. Mosq. Control Assoc.* **2:**44–47.
84. Reisen, W.K., 1986, Overwintering studies on *Culex tarsalis* (Diptera: Culicidae) in Kern County, California: Life stages sensitive to diapause induction cues. *Ann. Entomol. Soc. Am.* **79:**674–676.
85. Reisen, W.K., Meyer, R.O., and Milby, M.M., 1986, Overwintering studies on *Culex tarsalis* (Diptera: Culicidae) in Kern County, California: Survival and the experimental induction and termination of diapause, *Ann. Entomol. Soc. Am.* **79:**664–673.
86. Reisen, W.K., Meyer, R.P., and Milby, M.M., 1986, Overwintering studies on *Culex tarsalis* (Diptera: Culicidae) Kern County, California: Temporal changes in abundance and reproduction status with comparative observations on *C. quinquefasciatus* (Diptera: Culicidae). *Ann. Entomol. Soc. Am.* **79:**677–685.

87. Rosen, L., Shroyer, D.A., and Lien, J.H., 1980, Transovarial transmission of Japanese encephalitis virus by *Culex tritaeniorhynchus* mosquitoes, *Am. J. Trop. Med. Hyg.* **29:**711–712.
88. Roubaud, E., 1923, Les desharmonies de la fonction renale et leurs consequences biologiques chez les mosquitoes, *Ann. Inst. Pasteur* **37:**627–679.
89. Roubaud, E., 1982, Cycle autogene d'attente et generations hivernales suractives inapparentes chez les mostique commun *Culex pipiens* L., *C.R. Acad. Sci. Paris* **188:**735–738.
90. Rush, W.A., 1962, Observations on an overwintering population of *Culex tarsalis* with notes on other species, *Mosq. News* **22:**176–181.
91. Rush, W.A., Brennan, J.M., and Eklund, C.M., 1958, a natural hibernation site of the mosquito *Culex tarsalis* Coquillett in the Columbia River Basin, Washington, *Mosq. News* **18:**288–293.
92. Sanburg, L.L., and Larsen, J.R., 1973, The effect of photoperiod and temperature on ovarian development in *Culex pipiens pipiens* L., *J. Insect. Physiol.* **19:**1173–1190.
93. Schaefer, C.H., and Washino, R.K., 1969, Changes in the composition of lipids and fatty acids in adult *Culex tarsalis* and *Anopheles freeborni* during the overwintering period, *J. Insect Physiol.* **15:**395–402.
94. Schaefer, C.H., and Washino, R.K., 1974, Lipid contents of some overwintering adult mosquitoes collected from different parts of northern California, *Mosq. News* **34:**207–210.
95. Schaefer, C.H., Miura, T., and Washino, R.K., 1971, Studies on the overwintering biology of natural populations of *Anopheles freeborni* and *Culex tarsalis* in California, *Mosq. News* **31:**153–157.
96. Schlesinger, R.W., 1980, Virus–host interactions in natural and experimental infections with alphaviruses and flaviviruses, in: R.W. Schlesinger, (ed.), *The Togaviruses,* Academic Press, New York, pp. 83–106.
97. Schlesinger, R.W., Stollar, V., Igarashi, A., Guild, G.M., and Cleaves, G.R., 1978, Natural life cycle of arthropodborne togaviruses: Inferences from cell culture methods. in: E. Kurstak and K. Maramorosch (eds.), *Viruses and Environment,* Academic Press, New York, pp. 281–298.
98. Shemanchuk, J.A., 1965, On the hibernation of *Culex tarsalis* Coquillett, *Culiseta inornata* Williston, and *Anopheles earlei* Vargus (Diptera: Culicidae) in Alberta, *Mosq. News* **25:**456–462.
99. Skultab, S., and Eldridge, B.F., 1985, Ovarian diapause in *Culex peus* Speiser (Diptera: Culicidae), *J. Med. Entomol.* **22:**454–458.
100. Slaff, M.E., and Crans, W.J., 1977, Parous rates of overwintering *Culex pipiens pipiens* in New Jersey, *Mosq. News* **37:**11–14.
101. Slaff, M.E., and Crans, W.J., 1981, The activity and physiological status of pre- and post-hibernating *Culex salinarius* (Diptera: Culicidae) populations, *J. Med. Entomol.* **18:**65–68.
102. Spielman, A., 1964, Studies on autogeny in *Culex pipiens* populations in nature. I. Reproductive isolation between autogenous and anautogenous populations, *Am. J. Hyg.* **80:**175–183.
103. Spielman, A., 1971, Bionomics of autogenous mosquitoes, *Annu. Rev. Entomol.* **16:**231–248.
104. Spielman, A., 1974, Effect of synthetic juvenile hormone on ovarian diapause of *Culex pipiens* mosquitoes, *J. Med. Entomol.* **11:**223–225.

105. Spielman, A., and Wong, J., 1973, Environmental control of ovarian diapause in *Culex pipiens, Ann. Entomol. Soc. Am.* **66:**905–907.
106. Spielman, A., and Wong, J., 1973, Studies on autogeny in natural populations of *Culex pipiens*. III. Midsummer preparation for hibernation in autogenous populations, *J. Med. Entomol.* **10:**319–324.
107. Stollar, V., 1979, Defective interfering particles of togaviruses, *Curr. Top. Microbiol. Immunol.* **86:**35–66.
108. Sulaiman, S., and Service, M.W., 1983, Studies of hibernating populations of the mosquito *Culex pipiens* L. in southern and northern England, *J. Nat. Hist.* **17:**849–857.
109. Tate, P., and Vincent, M., 1936, The biology of autogenous and anautogenous races of *Culex pipiens* L. (Diptera: Culicidae), *Parasitology* **28:**115–145.
110. Tesh, R.B., and Schroyer, D.A., 1980, The mechanism of arbovirus transmission in mosquitoes: San Angelo virus in *Aedes albopictus, Am. J. Trop. Med. Hyg.* **29:**1394–1404.
111. Ura, M., 1976, Ecology of Japanese encephalitis virus in Okinawa, Japan. I. The investigation of pig and mosquito infection of the virus in Okinawa Island from 1966 to 1976. *Trop. Med. (Nagasaki)* **18:**151–163. [Cited in 82.].
112. Vinogradova, E.B., 1960, An experimental investigation of the ecological factors inducing imaginal diapause in bloodsucking mosquitoes (Diptera: Culicidae), *Entomol. Rev.* **39:**210–219.
113. Wallis, R.C., Taylor, R.M., McCollum, R.W., and Riordan, J.T., 1958. Study of hibernating mosquitoes in eastern equine encephalitis epidemic areas in Connecticut, *Mosq. News* **18:**1–4.
114. Washino, R.K., 1977, The physiological ecology of gonotropic dissociation and related phenomena in mosquitoes, *J. Med. Entomol.* **13:**381–388.
115. Watts, D.M., Pantuwatana, S., DeFoliart, G., Yuill, T. M., and Thompson, W.H., 1973, Transovarial transmission of LaCrosse virus (California encephalitis group) in the mosquito *Aedes triseriatus, Science* **182:**1140–11451.
116. Wesenberg-Lund, C., 1921, Contributions to the biology of the Danish Culicidae, *K. Dan. Vidensk. Selsk.* **7:**1–210.
117. Wilson, T.G., 1982. A correlation between juvenile hormone deficiency and vitellogenic degeneration in *Drosophila melanogaster, Roux's Arch. Dev. Biol.* **191:**257–263.
118. Wilton, D.P., and Smith, G.C., 1985, Ovarian diapause in three geographic strains of *Culex pipiens* (Diptera: Culicidae), *J. Med. Entomol.* **22:**524–528.

geographic range of a particular *Leishmania* species. The presence of *Le. braziliensis braziliensis,* was recently confirmed for the first time in Central America in Belize (33), where previously *Le. mexicana mexicana* was the only parasite known to cause human disease. These new findings may represent long-established, but undetected, foci, or reflect recent introductions of disease (e.g., possibly canine leishmaniasis in Oklahoma).

The existence of sympatric morphospecies of *Lutzomyia* further complicates epidemiological studies, though seemingly not to the extent as that shown by the asexual *Leishmania*. For example, females of *L. wellcomei,* a proven vector of *Le. b. braziliensis,* are difficult to distinguish from those of *L. complexa* in Brazil by morphological characters alone (84a, 131). There are additional examples in other species groups and it is likely that morphospecies or sibling species are more common in the genus than previously suspected. Few isozyme or allozyme analyses of *Lutzomyia* sand flies have been reported so far (16a, 123, 131, 164) but further studies using these methods and those based on monoclonal antibodies, cuticular hydrocarbons (144a) and genetic probes are expected to provide important information on inter- and intraspecific variation, especially as it relates to vector competence.

Transmission of leishmaniasis takes place in a variety of habitats in the New World, and results of epidemiological studies in one locality are not always applicable to another, even within similar habitats in the same country. This is especially true for the *Le. mexicana* complex and *Le. braziliensis* complex parasites. Major changes in the landscape, usually resulting from human activities, obviously affect species diversity and population densities of vectors and reservoirs. Ready *et al.* (132), studying adult sand flies, recently documented the effects of one such change in northern Brazil. With few exceptions (45, 140–143), little is known about the ecology of the immature stages of vectors; this is a serious handicap in terms of understanding population dynamics and for developing control strategies.

As a result, our discussion focuses mainly on adult sand flies and their role as vectors of leishmaniasis from the United States to Argentina where autochthonous human cases have been reported (178). We have drawn information from published and unpublished reports, emphasizing those that have appeared in the last decade and that are not readily available. There are several recent reviews on the leishmaniases (72, 73), including historical coverage of the American forms, a topic that we do not fully treat in the present chapter.

Classification of American Phlebotominae and *Leishmania*

Arguments for treating Phlebotominae as a family of Diptera (1, 2, 136) have been strengthened by a recent study (27) showing that the sper-

2
New World Vectors of the Leishmaniases

David G. Young and Phillip G. Lawyer

Introduction

The leishmaniases are a group of enzootic and zoonotic diseases caus by morphologically similar parasites in the genus *Leishmania* (Protozo Trypanosomatidae). Mammal reservoirs, of which there are many specie (76), may or may not show signs of infection (24, 52, 71). Furthermore some of the leishmanial species are host-specific and have not been re-ported in humans (e.g., *Le. hertigi* of porcupines). Those species that infect man cause an estimated 400,000 new cases each year throughout the world (178). Clinical symptoms of cutaneous, mucocutaneous, and visceral disease in man vary considerably, depending on the species of *Leishmania,* immunological responses of the individual, and other factors (178). Putative vectors of these diseases are sand flies in the genera *Lutzomyia* (New World) and *Phlebotomus* (Old World), but incrimination of specific vectors and mammal reservoirs remains undetermined in many foci (66).

One of the problems complicating epidemiological studies, particularly in the Americas, is that two or more leishmanial taxa and similar-appearing nonleishmanial trypanosomatids coexist in many localities (73). Moreover, the number of taxa continues to increase as more foci are studied (72). Identifying these organisms and associating them with vectors has been difficult, especially before isozyme analysis and other biochemical means of parasite identification became widely used within the last few years (28, 70, 108, 116, 157, 186). For example, leishmanial parasites were positively identified from only six of 812 sand flies found with natural flagellat infections in Panama (24).

New foci continue to be discovered, sometimes well beyond the know

David G. Young, Department of Entomology and Nematology, University of Florida, Gainesville, Florida 32611, USA.
Phillip G. Lawyer, Department of Entomology, Walter Reed Army Institute of Researcl Walter Reed Army Medical Center, Washington, D.C. 20307, USA.

matozoan ultrastructure of three *Phlebotomus* spp. differs from other psychodids in the subfamily Psychodinae. It will be necessary to study these structures in other subfamilies before conclusions can be drawn. The generic and subgeneric classification of Phlebotominae is even more controversial. Most American sand flies described before the mid-1960s were placed in the genus *Phlebotomus,* but Theodor in 1965 (163), as he did in 1948 (162), pointed out structural differences between species in this Old World genus and those of *Lutzomyia* in the New World. We recognize two other American genera in the subfamily, *Warileya* (six species) and *Brumptomyia* (24 species, including one undescribed species from South America). Sand flies in these smaller genera, unlike those in *Lutzomyia,* have not been implicated in disease transmission.

The 354 described and undescribed *Lutzomyia* spp. known to us are placed into 27 subgenera and equivalent species groups (100, 191). Some of these categories are recognized as genera by specialists who adopt Forattini's classification in whole or part (36, 37). It is difficult, however, to support the position of authors who recognize *Psychodopygus,* originally created as a subgenus, as a genus without simultaneously raising the rank of equally distinctive subgenera of *Lutzomyia* and of *Phlebotomus,* which are just as relevant to this discussion (100). Subgenera and species groups of *Lutzomyia* containing suspected or proven vectors of the leishmaniases are shown in Table 2.1.

Although the genus *Lutzomyia* is large and heterogeneous based on adult structure (163), it is not known if it is polyphyletic. A better knowledge of the morphology of immature stages, including larval polytene and brain cell chromosomes (69a, 177), may help set generic limits, but at present the larvae of only 54 *Lutzomyia* spp. have been described (e.g., 46, 169). Similarly, only a few studies on egg structure have been made (172, 199).

The infrageneric classification of *Leishmania* proposed by Lainson and Shaw (74, 75), with later modifications (72, 76), has become widely accepted. The species and subspecies are grouped into three taxonomic sections according to where they develop in their vectors. Thus, *Leishmania* in the Section Hypopylaria (two species infecting Old World lizards) multiply and develop only in the hindgut. Transmission from infected fly to lizard may be by ingestion (64), but this has not been demonstrated. Most *Leishmania* in the New World are placed either in the Section Peripylaria (*Le. braziliensis* complex) or in Suprapylaria (*Le. donovani, Le. hertigi,* and *Le. mexicana* complexes). The taxonomic position of *Le. herreri,* discovered in Costa Rican sloths (196), remains undetermined (72). Peripylarian leishmanias attach to the hindgut cuticle of sand flies, but they also move forward to the mid- and foreguts, eventually reaching the head region. The suprapylarian parasites do not attach to the hindgut wall, though slender promastigotes are sometimes observed moving freely in the lumen. Development and multiplication occur initially in the midgut, with later dispersal to the head region (see next section). *Leishmania henrici,* discovered in the blood and cloaca of *Anolis* lizards from Martinique,

TABLE 2.1. Suspected or proven vectors of leishmaniasis in the New World.[a]

Lutzomyia subgenus or species group	*Lutzomyia* species or subspecies	Associated *Leishmania*	Where suspected or incriminated as a vector	Reference
Subgenus *Lutzomyia*	*diabolica*	*Le. mexicana*	Mexico, Texas	86
	gomezi	*Le. b. panamensis*	Panama	24
	longipalis	*Le. d. chagasi*	Mexico to Argentina	84, 90
migonei species group	*migonei*	*Leishmania*	Brazil, Venezuela	76
Subgenus *Nyssomyia*	*anduzei*	*Le. b. guyanensis*	Brazil, Guyanas	5, 76
	flaviscutellata	*Le. m. amazonensis*	Brazil, Venezuela, Trinidad	73, 125, 165
	intermedia	Leishmania	Brazil	38, 129
	olmeca olmeca	*Le. m. mexicana*	Mexico, Belize, Guatemala	10, 182
	olmeca bicolor	*Le. b. aristedesi*	Panama	22
	olmeca nociva	*Le. m. amazonensis*	Brazil	J. Arias, personal communication
	trapidoi	*Le. b. panamensis*	Panama, Colombia, Costa Rica	24, 117, 195a
		Leishmania	Ecuador	47
	umbratilis	*Le. b. guyanensis*	Brazil, Guyanas, Colombia	76, 121, 193a
	whitmani	*Le. b. guyanensis*	Brazil, Guyanas	5, 76
		Leishmania	Brazil	76
	ylephiletor	*Le. b. panamensis*	Panama, Costa Rica	24, 194
		Le. m. mexicana	Guatemala	C. Porter, personal communication
Subgenus *Pintomyia*	*pessoai*	*Leishmania*	Brazil	38
Subgenus *Psychodopygus*	*ayrozai*	*Leishmania* of armadillos	Brazil	8
	hirsuta	*Leishmania*	Brazil	128a
	panamensis	*Le. b. panamensis*	Panama	24
		Leishmania	Venezuela	124
	paraensis	*Leishmania* of armadillos	Brazil	8
	wellcomei	*Le. b. braziliensis*	Brazil	73, 78
	yucumensis	*Le. b. braziliansis*	Bolivia	16a

TABLE 2.1. *continued*

Lutzomyia subgenus or species group	*Lutzomyia* species or subspecies	Associated *Leishmania*	Where suspected or incriminated as a vector	Reference
verrucarum species group	*christophei*	*Leishmania*	Dominican Republic	61
	spinicrassa	*Le. b. braziliensis*	Colombia	193a
	verrucarum	*Le. b. peruviana*	Peru	80
	sibling species of *townsendi*	*Le. m. garnhami*	Venezuela	4a, 151a, 153
vexator species group	*hartmanni*	*Leishmania*	Ecuador	47
	peruensis	*Le. b. peruviana*	Peru	51

[a]See text and other references (8, 47, 76, 144) for more details on these and other *Lutzomyia* species, some of which have been found naturally infected with unidentified promastigotes. Parasites listed here as *Leishmania* only are either unnamed or have not been specifically identified. Not all of the *Leishmania* listed here have been isolated from sand flies.

West Indies (87), may not belong in this genus (30). No other saurian *Leishmania* occurs with certainty in the Western Hemisphere.

Although our use of subspecific names for leishmanial taxa follows Lainson (72), it may not be entirely correct, because some subspecies are sympatric (73). It is expected that nomenclatural changes will soon be proposed, not only for this reason, but also because some of the named taxa may be junior synonyms of earlier described forms, notably those in the *Le. mexicana* complex (73, 116). In addition, not all parasites referred to as *Le. b. braziliensis* (in the broad sense) may belong in this taxon and new names may have to be created for them (73). Reference strains of some *Leishmania*, in effect, serve as neotypes and have added some stability to the classification of these parasites (178).

Development of *Leishmania* in Sand Flies

Experimental infections of *Leishmania* have been studied to varying degrees in more than 30 *Lutzomyia* spp. collected in nature or reared in the laboratory (4a, 17, 24, 61, 64, 76, 119, 122, 168a, 170). Improved methods for producing large numbers of sand flies in the laboratory (31, 67, 113, 118) have not only allowed investigators adequately to study these host/parasite relationships; they have also provided sufficient specimens for cross-breeding experiments (176), xenodiagnosis (19, 20, 24, 188), age grading, and other biological studies (103, 105, 130, 135).

Sand flies are pool feeders; the mouth parts and feeding mechanisms have been described (58, 96). The number of leishmanial amastigotes ingested during the blood meal may be critical for the successful establishment of infections within them. This varies according to the species of *Leishmania* and other factors (64). For example, Franke *et al.* (39) homogenized *L. longipalpis* females immediately after they had fed on hamsters experimentally infected with *Le. m. amazonensis* and *Le. b. panamensis*. The homogenates were then streaked onto the surface of blood agar plates and after incubating for 1–2 weeks, the isolated promastigote colonies were counted. Only 22–33% ($\underline{X}$ = 16%) of the flies that had fed on *Le. b. panamensis* ingested amastigotes (2–16/fly). In contrast, 23–100% ($\underline{X}$ = 61%) of flies that had engorged on *Le. m. amazonensis*-infected hamsters ingested leishmaniae (from 4 to >250/fly). The length of the sand fly mouth parts in relation to the distribution of amastigotes within the host's skin may also have an affect on the number of parasites ingested. The fascicle of a *Lutzomyia* species in the *verrucarum* group (not *L. townsendi*) ordinarily penetrates the skin of hamsters to a depth of 130 um, which is within the zone where amastigotes of *Le. m. garnhami* are especially abundant (17).

Subsequent development of *Leishmania* in sand flies was recently reviewed (64, 115) and only a brief summary will be given. Following the infecting blood meal, amastigotes probably divide one or more times before transforming into promastigotes in the midgut (64). Those of the *Le. braziliensis* subspecies invade the pylorus and, less frequently, the ileum of the hindgut. There, they transform into parasites (paramastigotes) having expanded, but shortened, flagella. This is a modification associated with the formation of hemidesmosomes within the flagellar sheath and serves to anchor the paramastigote to the hindgut wall, presumably a nutrient-rich site (64). Multiplication by binary fission continues to occur, but, eventually, free-living promastigotes are produced, which then move forward to the foregut and beyond, where they again appear as paramastigotes attached to the cuticular lining (64). Unlike the peripylarian parasites, those in the Section Suprapylaria grow mainly in the midgut and become attached to it at times, by inserting their unmodified flagella between microvilli or occasionally penetrating epithelial cells (115). Although there is no hindgut development, forward migration follows the same pattern as that of the peripylarian leishmanias (64). The rate of extrinsic development varies with temperature and species of *Leishmania*. Much work remains to be done on the physiology of sand flies in relation to leishmanial susceptibility, as influenced by digestion, peritropic membrane formation, lectins, and endosymbionts (63, 115).

Short-active promastigotes, believed to be infective forms (64, 145), may appear in the cibarium and mouth parts of the sand fly as early as 4 days following the infecting blood meal. It is assumed that these forms originate from paramastigotes attached to the stomodeal value, pharynx, or cibarium (86, 115), appearing when nutrients in the blood meal are depleted. Ingested sugars may block the attachment of paramastigotes at these sites, thus allowing the freely moving promastigotes to more easily invade the proboscis (115). If true, then the presence of sugars has an obvious effect on transmission. In addition, there is evidence that paramastigotes may interfere with labral and cibarial sensilla (65, 97), thus causing infective flies to probe more often than noninfected flies (9, 65). Further discussion of some of these observations is given by Bray (13).

Incrimination of Vectors

Criteria for incriminating vectors of disease have been discussed by many authors, including Killick-Kendrick and Ward (66), who applied these guidelines specifically to sand flies and the leishmaniases. Although there is general agreement on the validity of these criteria, the task of gathering evidence is often difficult.

Anthropophily

Because transmission of leishmaniasis to man normally occurs during probing or feeding of an infective sand fly, it is obvious that a vector must be anthropophilic to some degree. This is determined by direct observations (many reports in the literature) or less frequently by analysis of sand fly blood meals (23, 24, 76, 160, 161). About 90 of the 354 *Lutzomyia* species have been reported biting humans at one time or another, and of these, 27 species and subspecies are suspected or proven vectors (Table 2.1). More species will certainly be added to this list as additional studies are completed.

Leishmaniasis in humans or dogs has been reported from some localities where no mammal-feeding sand flies have been discovered [eg., Guadeloupe, West Indies (26) and Oklahoma (4)]. This may be due to limited vector search, to imported cases being erroneously considered as autochthonous, or to other, unknown reasons. In foci where only one anthropophilic sand fly species is present (e.g., *L. diabolica* in Texas and northern Mexico, *L. christophei* in the Dominican Republic), that species must be considered the likely vector to man, even before additional incriminating evidence is obtained. Even though two or more anthropophilic sand flies coexist in most endemic areas of the New World, the most common is not always the one that is responsible for transmission. Although *Lutzomyia flaviscutellata* is often cited as an infrequent man-biter, it is nevertheless, the only proven vector of *Le. m. amazonensis* to small mammals and man in northern Brazil (73). The incidence of human disease is understandably highest in foci where the vectors are strongly anthropophilic (e.g., *L. verrucarum* and perhaps *L. peruensis* in the Peruvian Andes). Other anthropophilic species, such as *L. davisi* in the Amazon basin, have not been implicated in transmission because the adults may have a short life span, their host range may exclude *Leishmania*-infected mammals, they may not be innately susceptible to infection, or they have been poorly studied.

Natural Leishmanial Infections in Sand Flies

Repeated isolations of *Leishmania,* indistinguishable from those causing human disease, from anthropophilic sand flies not only provide strong evidence for incriminating vector species (66), but also serve as a basis for estimating the risk of infection at a given time and place (91, 182). Traditionally, female sand flies are individually dissected and examined for promastigotes, which are later identified following *in vivo* or *in vitro* cultures (178). An overall infection rate is then calculated, but ideally a more realistic and significant value should be based on the proportion of parous females found infected. However, no long-term investigations of this kind have been carried out in the Americas.

Attempts to culture promastigotes are not always successful even when known leishmaniae observed in sand flies or mammals are inoculated into Syrian hamsters or suitable culture medium. The infective stage of the parasite may be absent (145) or the parasite itself may not be adaptable to these cultures (72).

Many of the difficulties associated with isolating leishmanias from their intrinsic or extrinsic hosts may soon be overcome by the use of newly developed immunoassays based on highly specific monoclonal antibodies. McMahon-Pratt *et al.* (109) and Frankenburg *et al.* (40) demonstrated that promastigotes from experimentally infected flies can be rapidly and accurately identified by these techniques. It is not necessary to culture *Leishmania* for these determinations and it is likely that promastigotes in dry preserved sand flies can be identified, as was demonstrated for *Plasmodium* sporozoites in mosquitoes (15). Furthermore, it should be possible to identify sand flies, especially females of sibling species, using these or similar methods.

Meanwhile, preserving field-collected sand flies in liquid nitrogen is an excellent, but little-used, method for holding specimens until they can be dissected in the laboratory, usually under better conditions (112, 117, 193a). Promastigotes in sand flies remain viable for 26 months (193a), and probably for years, when held at −196°C. Freezing sand flies in the field allows more time for collections and observations, employing time otherwise spent dissecting, and reduces or eliminates the need for transporting hamsters, culture media, and dissecting equipment to remote field sites. In this regard, light-weight cryogenic tanks, lined inside with adsorbent material to prevent the spilling of liquid nitrogen, are particularly useful.

Triturating groups of sand flies to recover *Leishmania* has been done in some studies (22, 51, 73) and saves time when the only objective is to recover parasites. Yet, there is a danger of mixing two or more species in a single pool; only an estimated infection rate can be calculated; and no information can be obtained on the age structure of the sand fly populations using present methods, i.e., follicular dilatations of individual females (93, 99, 103, 135).

Demonstration of Transmission by Sand Flies and Other Arthropods

Stable flies or dog flies, *Stomoxys calcitrans*, can mechanically transmit *Le. mexicana* (and probably other *Leishmania* as well) from hamster to hamster under experimental conditions (77), but this mode of transmission of *Le. mexicana* is assumed to be rare or nonexistent in nature. These insects feed readily on dogs and their interrupted feeding behavior (16) would favor mechanical transmission of some of the *Le. donovani* complex parasites, which are often abundant in the skin of infected dogs (178).

Sherlock (156) studied a focus of *Le. d. chagasi* in northeastern Brazil, concluding that the brown dog tick, *Rhipicephalus sanguineus*, transmitted the parasite to dogs at localities where *L. longipalpis*, the proven sand fly vector, was absent. However, promastigotes found in ticks were not conclusively identified as *Leishmania*. Recently, McKenzie (107) observed that *Leishmania (Le. donovani infantum?)* in dogs from Oklahoma and Kansas could survive in experimentally infected ticks of the same species for up to 160 days after their infecting blood meals, and for as long as 130 days postmolting (nymph to adult). She also demonstrated that adult ticks experimentally fed as nymphs on infected dogs about 30 days earlier transmitted the parasite by bite to another dog (see p. 50). Although these results contradict those of another tick transmission trial (137), additional studies are needed to fully assess the role, if any, that ticks may play in the natural transmission of some *Le. donovani* complex parasites.

In general, however, sand flies are regarded as the only important vectors of the leishmaniases. It is apparent under experimental conditions that the *Le. mexicana* complex parasites can develop in and be transmitted by many *Lutzomyia* species, some of which are not natural vectors (185). This lack of host specificity is perhaps analogous to *Trypanosoma cruzi* and triatomine bugs; while nearly all of these are potential vectors, only those bugs having certain habits and geographic distributions serve as actual vectors of the disease to man (88). Examples of *Lutzomyia* females proven to experimentally transmit *Le. mexicana* or closely related parasites by bite include: *L. anthophora* (R. Endris and D. Young, unpublished data); *L. christophei* (61); *L. cruciata* (181); *L. diabolica* (86); *L. flaviscutellata* (175); *L. longipalpis* (25); *L. panamensis* or ally (158, 184); *L. renei* (25); *L. shannoni* (P. Lawyer and D. Young, unpublished data), and *L.* sp. near *townsendi* (151a). Only *L. flaviscutellata* and *L. o. olmeca* are proven vectors of any of *Le. mexicana* parasites. The other listed sand flies are suspected vectors (Table 2.1) or else do not appear to be involved in natural transmission cycles. *Leishmania d. chagasi*, also in the Section Suprapylaria, was experimentally transmitted to hamsters by laboratory-bred females of *L. longipalpis* (79). This species was previously incriminated as the natural vector on epidemiological grounds (biting habits, general concordance of geographic distribution with that of the parasite, and other data). Recently, promastigotes from wild-caught *L. longipalpis* in Brazil (84) and Bolivia (90) were positively identified as *Le. d. chagasi,* thus completing the chain of evidence needed to establish its position as the principal vector of New World visceral leishmaniasis.

Development of the peripylarian leishmanias (*Le. braziliensis* complex parasites) in sand flies has also been studied, as discussed previously, but there are no reports of successful experimental transmissions. Attempts to infect *L. flaviscutellata,* an unnatural vector of these parasites, have failed (73), and probably other, unreported attempts with different *Lutzomyia* species have been equally discouraging. This may be due in part to the difficulty in colonizing peripylarian vectors in the laboratory (73), or more likely because it is difficult to infect high proportions of sand flies with some of these parasites under experimental conditions (39).

Documentation of naturally infected, identified sand flies transmitting leishmaniasis to man adds further weight to the incrimination of vector species, but such reports only a few isolated observations. Pifano *et al.* (124) incriminated *L. panamensis* as a vector in Venezuela when a collector contracted leishmaniasis (*Leishmania* species undetermined) 4 weeks after being bitten by 17 wild-caught *L. panamensis* females. Biagi *et al.* (10) reported that a field-collected female of *L. o. olmeca* (as *Phlebotomus flaviscutellatus*) transmitted *Le. m. mexicana* to a human volunteer in Mexico. Additional evidence from there and Belize (182) supported the view that *L. o. olmeca* is an important natural vector in these areas.

Other Epidemiological Evidence

Lutzomyia species closely related to proven vectors and with similar host preferences are sometimes presumed to be vectors in the absence of other evidence. For example, *Lutzomyia o. bicolor* has not been found naturally infected with *Le. m. aristedesi,* but females of this subspecies probably transmit this parasite to rodents, especially *Orozomys,* at Sasardi, Panama, where these animals are commonly infected (54). Like its close relatives in the *flaviscutellata* complex, *L. o. bicolor* is readily attracted to rodents (22).

This subspecies and many other sand flies are not known to transmit *Leishmania* to man, yet some of them may help maintain infections among reservoir hosts. However, these enzootic associations and the extent to which they occur have been poorly studied in the Americas. Christensen *et al.* (24) showed that *L. shannoni* in Panama feeds frequently on sloths and suggested that this sand fly may serve as a vector of *Le. b. panamensis* among these animals, but not to man.

Other observations providing clues to the incrimination of vector species obviously include knowing where and when transmission takes place and which age groups of people are most infected. Knowledge of the habits of people in relation to those of man-biting flies can provide important information in this regard. Ward *et al.* (174) first reported that *L. wellcomei* was an avid day biter at Serra das Carajás, Pará State, Brazil, where the *Leishmania* infection rate was high among men working in forests during the day. Subsequent isolations of *Le. b. braziliensis* from *L. wellcomei*, along with other information (73), incriminated this species as the principal vector.

Geographic Distribution of Vectors

In contrast to the *Phlebotomus* species in the Old World, which live mostly in temperate zones, the *Lutzomyia* sand flies are essentially tropical insects (94). The few species occurring in North America were probably derived from such stock (192). The present center of distribution of the *Lutzomyia* species lies in the lowland forests of South America, east of the Andes, where 40 or more species may coexist in a single locality (193). One explanation for this rich diversity is that recurring dry periods during the Pleistocene epoch, and probably before, served to isolate conspecific populations of sand flies in wet refugia (190). In time, some of these populations became reproductively isolated. Alternating wet periods allowed sand flies to expand their ranges, some of which apparently spread around, not across, the Andes Mountains to colonize areas mostly to the west and north.

Oceanic islands, including those of the West Indies (excluding Trinidad, which is a continental island) have depauperate sand fly faunas, probably because of the poor flight capacity of sand flies (35). Most species feed on lizards, but a few, such as *L. christophei,* a suspected vector of leishmaniasis in the Dominican Republic, and *L. orestes,* a closely related species known to bite humans in Cuba (110), are mammalophilic. Other sand flies, now extinct, once inhabited Hispaniola. Examination of 14 fossil sand flies of two undescribed species (D. Young and R. Johnson, unpublished data) preserved in Dominican amber and believed to be about 26 million years old (146), showed that they closely resemble modern forms (see Fig. 2.1) (127).

On continental land masses in the neotropics, the sand fly fauna is much richer. However, no sand flies have been reported from Chile (106) and only two species are known from Uruguay (106). Imported human cases of leishmaniases were recently reported from Chile (147).

Of the proven vectors of leishmaniases, *L. longipalpis* has the widest geographic range (disjunct distribution in dry areas from southern Mexico to northern Argentina), but there is some evidence indicating that it represents a species complex consisting of two or more taxa (176). The same may be true for *L. shannoni,* which occurs from Maryland to northern Argentina, but there are no data to support this hypothesis. Some other species, less widespread, have markedly disjunct distributions. An undescribed *Lutzomyia* species, previously misidentified as *L. townsendi* in western Venezuela (D. Feliciangeli, J. Murillo, and D. Young, unpublished

FIGURE 2.1. An unnamed *Lutzomyia* female preserved in amber from the Dominican Republic and believed to be ~26 million years old.

data), and a suspected vector there (153), was recently discovered in a coffee-growing area of Costa Rica where *Le. b. panamensis* occurs (R. Zeledón, personal communication). This sand fly has not been reported elsewhere. Further information on the geographic distributions of individual vectors is given later in this chapter.

Some Habits of Sand Flies as Related to Leishmaniasis Transmission

Lewis (94) reviewed this broad subject; here we discuss some aspects of feeding, longevity, and flight range, while later giving more information on habits of specific vectors. Killick-Kendrick (63) discussed some aspects of the biology of phlebotomines that have not been well studied.

Feeding

Although it has not been established where or when adult sand flies obtain sugar meals in nature, it is assumed that both sexes require carbohydrates for energy and longevity (98). Obviously, both factors have an important bearing on the transmission of leishmaniases. The presence of fructose, glucose, and sucrose has been detected in the crops of Old and New World phlebotomines (64, 98).

Williams (182) examined 2466 individual sand flies of eight species in Belize, noting that the crops of 155 contained varying amounts of a clear fluid, probably sugar solutions as was determined previously (98). J. Alexander (personal communication) failed to find pollen grains on the heads or bodies of 2371 Panamanian sand flies examined and, therefore, concluded that floral nectar does not serve as a usual sugar source for these insects. Other investigators have drawn the same conclusion and have suggested that extrafloral nectaries, honeydew, ripe fruits, and other plant juices may serve as natural sugar sources. Recent observations that males and females of *Phlebotomus papatasi* pierce stems and leaves of plants under experimental conditions to obtain sugars suggest that this behavior may pertain to New World sand flies as well (148a).

Alexander analyzed 554 Panamanian sand flies using the cold anthrone test (166) and determined that fructose, at least, was present in 129 males and females of six *Lutzomyia* species. The high proportion (58%) of 65 males collected from tree trunks and having such sugar may reflect the possibility that they, like male mosquitoes, require more sugar meals than females because they do not accumulate fat as adults (167). However, up to 37% of nongravid females from tree trunk collections had ingested detectable levels of sugar.

Ingested sugars are stored either in the flexible crop or pass directly to the midgut of the fly, depending on the depth of penetration of the mouthparts and interaction of the palpus with the surface of the sugar source (130, 148). Some observations indicate that the presence of ingested sugar may help stimulate host-seeking of female flies (182).

In addition to sugars, most *Lutzomyia* females require one or more blood meals to initiate ovarian development. There is a range of gonotrophic patterns that varies within and among sand fly populations. Johnson (59) studied an obligatory autogenous subpopulation of *L. gomezi* in Panama, noting that females did not require a blood meal for the maturation of the first batch of eggs. Lewis (93) and Perkins (122) observed the same phenomenon in a closely related species, *L. cruciata,* in Belize and Florida, respectively.

Facultative autogeny has been observed in a proportion of *L. beltrani* in Belize (183) and *L. shannoni* females from Florida (122). Nearly all (96%) of 349 *L. shannoni* reared in the laboratory deposited up to 40 eggs within 4 days of emergence without having had blood meals. Yet, other females, containing developing eggs from the same laboratory colony, readily took multiple blood meals prior to the first oviposition. Such gonotrophic disconcordance, also observed in anautogenous species (14), obviously enhances vector potential, but the extent to which multiple feedings occur during a single ovarian cycle in nature is unknown. The typical gonotrophic pattern, anautogeny, is exhibited by most *Lutzomyia* vectors, which require one or more blood meals for the development of all their eggs.

Host-seeking behavior is complex, but little-studied in sand flies. The various, but not necessarily separate, events leading to the ingestion of vertebrate blood include activation, orientation, landing, and feeding (62, p. 124). Most anautogenous females are physiologically ready to feed within 1–4 days after emerging from the pupal stage (32, 67). Suitable combinations of light intensity, ambient temperature, relative humidity, air movement, and other physical conditions stimulate hungry sand flies to search for blood meals.

This behavior varies according to phlebotomine species, vertebrate host, and habitat (126). For nest-inhabiting species such as *L. anthophora* in Mexico and Texas, the search may be short because their hosts occupy the same microhabitat and blood feeding corresponds to the time when the host is present and probably resting (189). Females of *L. anthophora* are unable to fly after a full blood meal and there seems to be no selection pressure for them to do otherwise.

In contrast, most other *Lutzomyia* species must travel greater distances from resting sites to find their hosts. Forest species, such as *L. trapidoi* and *L. umbratilis,* search for arboreal mammals by moving in short, hopping flights along tree trunks and branches (21). In the same habitat, other vectors (e.g., those in the *L. flaviscutellata* complex) restrict their hunting mostly to sites at or near ground level. *Lutzomyia longipalpis,* a robust sand fly, commonly occurs in treeless localities and is not closely associated with mammal nests. Its flight range has not been determined, but it is likely that both sexes fly considerable distances to locate their hosts. This species and others, such as *L. gomezi, L. diabolica, L. intermedia, L. verrucarum, L. spinicrassa,* and *L. panamensis,* will readily enter houses to bite the occupants (many scattered references). Yet, in general, these and other vectors are exophilic; there are no truly domestic sand flies in the Americas.

In Panama, Christensen *et al.* (21) found that a high proportion of gravid *Lutzo nyia* females were attracted to caged bait animals, besides man. They suggested that blood meals taken shortly before oviposition could provide an energy source that would lessen the deleterious effects associated with egg laying. However, there was no evidence that these females actually took blood meals (due to the design of the baited traps), and thus the possibility remains that some of them, at least, may have been attracted to conspecific males resting near the animals. Results of other studies in the neotropical forests indicate that the proportion of gravid females coming to human bait is less than 10% (D. Young, unpublished data).

Orientation to a host probably involves several stimuli, including host odors (130), temperature gradients (149), and, perhaps, pheromones released from feeding females or conspecific males that for some species are also attracted to vertebrates (111). Schlein *et al.* (149) demonstrated that *Phlebotomus papatasi* females secrete an aggregation pheromone, probably from palpal glands, while taking a blood meal. The response to this volatile pheromone, not yet reported in American sand flies, varied according to the temperature of the experimental feeding chamber.

Males of *L. longipalpis* normally outnumber females on or near their vertebrate hosts (120, 198) and their vigorous wing beating behavior may promote better dispersal of sexual pheromones, believed to be secreted from dorsal abdominal glands (85). Such glands have not been reported in other sand flies, but rapid wing beating seems to be a common premating trait among all male flies (11).

The size and structure of sand fly mouth parts can provide clues regarding host preferences. Lewis (96) observed that mammal feeders generally have a long labrum and that it and the paired maxillae possess hooked teeth, rather than ridge-tipped maxillae, which are associated with reptile feeders. Most American Phlebotominae are mammal feeders and, with few exceptions, are regarded as opportunistic, feeding on a variety of mammals, depending largely on their availability (21, 160, 161). Precipitin testing of sand fly blood meals has provided important epidemiological information on vector host preferences in Brazil (23) and Panama (160, 161), but has been little used elsewhere in the neotropics. Another technique for studying this subject involves inoculating individual vertebrates with rubidium, a relatively stable, nontoxic element that was detected in mosquitoes 6 days after they had fed on experimentally marked birds (69). If this method proves applicable to phlebotomines and mammals, it may supplement precipitin testing of blood meals and additionally provided data on the dispersal and longevity of sand flies after they have fed.

Longevity and Flight Range

We know little about longevity of sand fly vectors in nature and it is difficult to study this subject there or in the laboratory, where it is impossible to duplicate natural conditions. Nevertheless, studies of laboratory-bred phlebotomines have provided some understanding of longevity.

In terms of their survival, and therefore that of associated leishmanias, sand flies have evolved mechanisms for coping with periods of dryness or cold temperatures, that would reduce or perhaps eliminate their populations. These unfavorable periods are not always predictable, even in the neotropics, and thus no single strategy would enable all individuals to survive (86). Females of some *Lutzomyia* species occurring in both the Neotropical and Nearctic regions lay normal-developing and quiescent or diapausing eggs in the same batch (60, 86). This increases the probability that some of the progeny will eclose and live under favorable conditions. This adaptation has also been observed in other insects (168); the type of egg produced is dependent on the temperature and photoperiod (and relative humidity levels?) experienced by the parent during the preoviposition period (168). Lawyer (86) studied the effects of day length on adult *L. diabolica* from Texas in relation to the proportion of diapausing eggs laid by females (see p. 52). Although the influence of photoperiod may be less significant in the neotropics, Johnson and Hertig (60) observed uneven hatching rates within single egg batches of *L. gomezi, L. panamensis,* and *L. geniculata* when kept under identical conditions. It is also assumed that larvae of some neotropical species respond to unfavorable conditions by becoming dormant, but this may not represent true diapause.

From numerous laboratory studies (eg., 60), we know that most adult females of *Lutzomyia* generally live from 1 to 4 weeks, depending on species, temperature, and other factors. There is virtually no information on the adult life span of sand flies in nature. Chaniotis *et al.* (18) recovered five marked males of two Panamanian species 15 days after they had been dusted with fluorescent powder, but obviously their ages were unknown before they were marked nor was it known how long they would have lived if not captured. It has been assumed that high female mortality associated with oviposition of laboratory bred flies does not occur in nature (64); however, this was not supported by one age grading study of field-collected *Phlebotomus papatasi* (103).

Similarly, the flight ranges of sand flies are poorly known. Chaniotis *et al.* (18) marked about 20,000 sand flies in a Panamanian tropical forest. Subsequent recapture of marked flies showed that most (~90%) dispersed within 57 m of the release point. The maximum distance traveled by a single fly in one day was 200 m, the farthest distance monitored from the release point.

Lutzomyia Vectors and Associated Leishmania

Leishmania herreri, a parasite of sloths in Costa Rica (196), is not considered further because of its uncertain taxonomic status (72). *Lutzomyia trapidoi, L. ylephiletor,* and *L. shannoni* are implicated as vectors (196).

Section Peripylaria (*Leishmania braziliensis* Complex)

Most of the peripylarian leishmanias cause single or multiple ulcerated skin lesions in humans. Mucocutaneous leishmaniasis, due to *Le. b. braziliensis,* and *Le. b. panamensis* (147a) is a serious disease that may evolve months or years after the healing of the primary lesion(s) found elsewhere on the body (178).

Leishmania Braziliensis Braziliensis

Lainson (73) noted that human isolates, referred to as *Le. b. braziliensis,* from the Amazonian lowlands may actually consist of several distinct leishmanias. The reference strain (not a topotype) of this parasite was isolated from a patient at Serra dos Carajás, Pará, Brazil, in general a relatively high, well-drained locality (134). There, Lainson *et al.* (78) found promastigotes in females of *L. paraensis* (2 positive of 175 examined), *L. amazonensis* (1 of 127), and *L. wellcomei* (3 of 1656); however, only those recovered from the latter species were identified as *Le. b. braziliensis.*

Ward *et al.* (174), as noted earlier, observed that *L. wellcomei* females avidly bit man during the day and night. Other specimens were captured in rodent-baited Disney traps, but not on tree trunks or in animal burrows. The diurnal resting sites and host preference range of this species remain unknown.

Significantly, Wilkes *et al.* (180) showed that a high proportion (51%) of females collected during the day at this locality were parous and were thus potentially infective. Night collections yielded a lower proportion of parous flies (25%). Their findings supported laboratory observations showing that *L. wellcomei* females ordinarily lay their eggs at night, after which they seek another blood meal within 24 hr (180). Autogeny in *L. wellcomei* has not been observed and the females are gonotrophically concordant.

Populations of *L. wellcomei,* the only proven vector of *Le. b. braziliensis* known, decrease during the dry season at Serra dos Carajás (July–September), a time corresponding to decreased human infection (134). This species also occurs in the Brazilian states of Ceará (133) and Amazonas, near the Rio Urubu (J. Arias and D. Young, unpublished data). However, *Le. b. braziliensis* occurs outside the geographic range of *L. wellcomei* and thus other vectors must be involved with transmission. Among these suspected vectors (76), *L. intermedia, L. pessoai, L. whitmani,* and others have been implicated in transmission in parts of southern Brazil (37, 38, 72, 129), *L. panamensis* in Venezuela (3, 124), and *L. yucumensis* in Bolivia (16a). The latter species also occurs in foci of mucocutaneous leishmaniasis in Peru [as dark form of *L. c. carrerai* (193)] and in parts of Amazonia Brazil (J. Arias, personal communication).

Recently, this parasite was isolated and identified biochemically (R. Kreutzer, personal communication) from patients living near or in coffee groves at Arboledas, Norte de Santander, Colombia, where *L. spinicrassa* (*verrucarum* group) and *L. gomezi* (subgenus *Lutzomyia*) accounted for 79% of all sand flies collected in man-biting collections from April 1984 to August 1986 (193a). Of 10,086 *L. spinicrassa* females dissected, four contained promastigotes. One infection was later identified as *Le. b. braziliensis* (193a). Promastigotes in two of the other *L. spinicrassa* females were attached to the pylorus and were also seen in the mid- and foreguts. This suggests that they may also represent *Le. b. braziliensis* rather than *Endotrypanum* parasites (154) of sloths, because these animals are absent at this locality.

Vector information on *Le. b. braziliensis* in Belize (33), Ecuador (139), and Paraguay (171) is virtually nonexistant.

Leishmania Braziliensis Guyanensis

The epidemiology of leishmaniasis caused by *Le. b. guyanensis* is relatively well understood in northern Brazil (5, 81, 83, 132) and French Guiana (41, 89, 91, 92, 121). The putative vector is *L. umbratilis* (134a) a forest-dwelling sand fly, often abundant in the canopy at the end and beginning of the rainy seasons (91). This species normally rests on tree trunks and will attack man during the day and at night (76). Its usual hosts are arboreal mammals, especially sloths *(Choloepus didactylus)* and anteaters *(Tamandua tetradactyla),* which serve as reservoirs of this disease. Opossums *(Didelphis marsupialis)* have also been found naturally infected at Manaus, Brazil (6, 7), but they may represent accidental, "dead end hosts" (73).

Infection rates as high as 15.9% have been recorded in wild-caught *L. umbratilis* in French Guiana and there, at certain places and times of the year, it has been calculated that a person will be bitten by at least one infected *L. umbratilis* per hour (91). Susceptible persons entering endemic forests are at high risk; presently in Manaus, from 20 to 30 new cases of leishmaniasis due to this parasite are seen daily in clinics (J. Arias, personal communication). Secondary vectors, also in the subgenus *Nyssomyia,* include *L. whitmani* and *L. anduzei* (5, 81).

Lutzomyia umbratilis also occurs in the Amazon basin of Colombia (190), Surinam (56a, 179), Venezuela in Bolivar State (D. Feliciangeli, personal communication), and Peru (101). Human infections of *Le. b. guyanensis* have been reported in Surinam (179) and more recently in southeastern Colombia (R. Kreutzer, personal communication). One *L. umbratilis,* collected near Leticia, had a natural infection of *Le. b. guyanensis* (193a).

Leishmania Braziliensis Panamensis

Considerable information on the epidemiology of leishmaniasis caused by this parasite has been gathered in Panama during the past 35 years. Christensen *et al.* (24) reviewed this subject, giving references to the many detailed studies on the distribution, ecology, and disease relationships of Panamanian sand flies.

The epidemiologies of leishmaniasis due to this subspecies and *Le. b. guyanensis* are remarkably similar (73). Once again, arboreal mammals, especially sloths that have cryptic infections of *Le. b. panamensis* (52, 55), serve as principal reservoirs and *Lutzomyia (Nyssomyia)* species are the main vectors, notably *L. trapidoi* and *L. ylephiletor.* These species have been found naturally infected in Panama, as have *L. gomezi* and *L. panamensis* (24), but their role in transmission is less clear and their geographic ranges are wider than those of the two *Nyssomyia* species (106).

Leishmania braziliensis panamensis also occurs in Honduras (197), Costa Rica (194, 195), Colombia (117; R. Kreutzer, personal communication), and probably Ecuador, where Hashiguchi *et al.* (47) recently isolated *Le. braziliensis* complex parasites from females of *L. trapidoi* and *L. hartmanni.* The latter species was previously unsuspected as a vector of any of the leishmaniases. Elsewhere, natural infections believed or proven to be *Le. b. panamensis* have been reported in *L. ylephiletor* in Costa Rica (194) and *L. trapidoi* in Colombia (193a) and Costa Rica (195a).

In Central America, from Panama to Guatemala, the previously unsuspected occurrence of *L. edentula,* an anthropophilic sibling species of these *Nyssomyia* spp., must be considered when studying vectors in this region (D. Young and C. Porter, unpublished data). In fact, one of the male paratypes of "*L. ylephiletor*" from Chiriqui Province, Panama represents not that species, but *L. edentula*. The female characteristically has nine or more spermathecal annuli and very wide sperm ducts, unlike those of *L. trapidoi* or *L. ylephiletor*. It is not known whether *L. edentula* or the undescribed *verrucarum* group species in Costa Rica mentioned earlier are involved in the transmission of *Le. b. panamensis* or other *Leishmania*.

Leishmania Braziliensis Peruviana

Lainson *et al.* (80) reviewed information on the epidemiology of *Le. b. peruviana* in Peru and demonstrated for the first time that it is a typical peripylarian *Leishmania* based on its growth pattern in experimentally infected *L. longipalpis*. This parasite occurs in the western slopes and valleys of the Peruvian Andes, where fewer than six phlebotomine species have been reported (56, 193). In Argentina, the presence of *Le. b. peruviana* has not been confirmed; no recent isolates from there have been identified as *Le. b. peruviana* and the sand fly fauna of that country is poorly known (106).

In Peru, where leishmaniasis ("uta") has been linked with sand flies for centuries (53), there are still many unanswered questions about natural transmission cycles. Herrer (51) inoculated 97 triturated *L. peruensis* females into seven hamsters, two of which subsequently became infected with leishmaniasis (no other natural infections of sand flies have been reported in Peru). The altitudinal distribution of *Le. b. peruviana* in the western Andes, from approximately 900 to 3000 m above sea level (80), does not completely coincide with that of *L. peruensis,* which is rare or absent below 1400 m (56). *Lutzomyia verrucarum,* on the other hand, is also a suspected vector because of its wide altitudinal range, abundance, peridomestic habits, and aggressive man-biting behavior (56). Furthermore, another sand fly, *L. oligodonta,* which may also be anthropophilic, was recently discovered near Cocachacra, Peru, an endemic site for leishmaniasis in the Rimac Valley (193). It is possible, of course, that more than one vector species transmits *Le. b. peruviana* in these areas.

Although domestic dogs have been found naturally infected, and sympatric foxes have been experimentally infected with this parasite (48), Lainson (73) suggested that further attempts to find *Leishmania* in other mammals, especially rodents, are indicated. Herrer (50) reported that 21 of 47 sentinel hamsters placed in caves in the Santa Eulalia Valley of Peru and exposed to sand flies developed leishmanial lesions following 8 weeks of exposure. Near Trujillo, Peru, *Leishmania* was recently isolated from rats, *Rattus rattus* (L. Cruzado, personal communication), but these parasites may not be consubspecific with *Le. b. peruviana,* based on their rapid growth in hamsters. If rodents are subsequently incriminated as reservoirs, then the role of *L. noguchii* as a vector among them, but not to man, must be considered.

Leishmania Species from Armadillos

Lainson *et al.* (81) isolated *Leishmania* from blood, liver, and spleen of 3 of 14 armadillos *(Dasypus novemcinctus)* collected in north Pará State, Brazil. No skin lesions were observed. Among other observations, the hindgut as well as midgut development of these stocks in experimentally infected *L. longipalpis* showed that they belonged in the *Le. braziliensis* complex (81). Further characterization by enzyme electrophoresis demonstrated that the armadillo *Leishmania* differed from other members of this complex.

From wild-caught females of *L. ayrozai* and *L. paraensis* in Rondonia State, Brazil, Arias *et al.* (8) isolated and identified seven strains of this parasite by enzyme electrophoresis. These anthropophilic sand flies, belonging in the subgenus *Psychodopygus,* are widespread in the Amazon basin, but they have not been implicated in leishmaniasis transmission to humans.

Section Suprapylaria (*Leishmania donovani* Complex)

Leishmania Donovani Chagasi

There is unresolved controversy regarding the identity of this parasite, the etiological agent of American visceral leishmaniasis. Some authors suspect that some strains represent introduced *Le. d. infantum* from the Old World (68) because their enzyme profiles are essentially indistinguishable (82, 150); dogs serve as domestic reservoirs, and the incidence of infection is highest in children (178). In contrast, Jackson *et al.* (57, 73) distinguished *Le. d. infantum* from *Le. d. chagasi* by radiorespirometry and DNA analysis. Furthermore, it is argued that the widespread geographic distribution of *Le. d. chagasi* from Mexico to Argentina (171) indicates that it is a precinctive parasite in the Americas. An exception may be the *Leishmania* parasite causing canine leishmaniasis in Oklahoma and Kansas (4, 107).

The role of sylvatic mammals as additional reservoir hosts has not been fully determined, but, when better understood, may help settle this controversy. Foxes (*Cerdocyon* and *Lycalopex*) in Brazil (76) and the black rat *(Rattus rattus)* in Honduras (73) have been found naturally infected with *Leishmania,* probably *Le. d. chagasi.*

With few exceptions (26, 107), visceral leishmaniasis due to this parasite in the Western Hemisphere is associated with *L. longipalpis,* the habits and disease relationships of which were reviewed by Lainson and Shaw (76). This species, a proven vector as noted earlier, occurs in semiarid, often treeless areas from Mexico to Argentina (106). No infections of *Le. d. chagasi* have been found in Costa Rica (198) or in Panama (24). Males of *L. longipalpis* from Ceará State, Brazil and elsewhere (176) have two pale spots on abdominal tergites 3 and 4; those from most other localities have but one such spot on tergite 4. These probably represent the surface structure of abdominal glands that emit pheromones (85). Because of this difference and because cross-mating experiments (176) indicated that the two forms were reproductively isolated, in Brazil, at least, *L. longipalpis* apparently represents a species complex. One of us (D.G.Y.) examined males of *L. longipalpis* from Mexico, El Salvador, Costa Rica, Panama, Colombia, Bolivia, Venezuela, and Paraguay and noted that only those from the latter country ($N = 23$ from Asuncion) represented the two-spot form. According to Ward *et al.* (176) the two-spot form may be more domestic and anthropophilic than its counterpart. In any event, females of the one-spot form transmit *Le. d. chagasi* in many foci.

Lutzomyia longipalpis does not occur at Edmond, Oklahoma or Stanley, Kansas, where autochthonous leishmaniasis in foxhounds have been reported (4, 107; J. Fox, personal communication). Because *Leishmania* isolated from infected dogs has been identified as *Le. d. infantum* (28) and as *L. mexicana* (70), it is not known whether one or more leishmanial taxa are involved. The disease in dogs is generally consistent with that of visceral leishmaniasis (4), but some differences have been noted (107). Additional information, provided below, was kindly provided by J. Fox (personal communication) and McKenzie (107).

Since 1979, when the first four dogs from Oklahoma were diagnosed as having leishmaniasis (4), at least ten others have been found infected. One of these was shipped to the Edmond kennel from Stanley, Kansas. Leishmanial amastigotes were observed in this dog within 2 weeks after its arrival in Oklahoma, indicating that it contracted the disease in Kansas. The other infected dogs were born and raised at the kennel in Edmond. Continuous removal of infected dogs has not stopped transmission of this disease.

So far, only *L. vexator,* a reptile-feeding sand fly, has been reported at the Edmond site (192), but no adequate entomological survey has been carried out there or in Kansas, where no *Lutzomyia* species have been reported. The apparent lack of a suitable sand fly vector lead McKenzie to study the possibility that the brown dog tick, *Rhipicephalus sanguineus,* could serve as a biological vector.

Accordingly, she allowed 500 laboratory-bred nymphs of this species to fully engorge on each of two naturally infected dogs: the one that had contracted the infection in Kansas, and another from the Edmond kennel. Before and after molting to the adult stage about 30 days later, some of the ticks were examined periodically for leishmanial infections by inoculating triturated tick guts into suitable culture medium. Positive cultures from ticks that had fed on the Kansas-infected dog 160 days earlier were recorded, thus demonstrating transstadial transmission for the first time. McKenzie, however, did not microscopically observe *Leishmania* in serial sections of tick guts, nor did she determine if they multiplied in ticks following their infecting blood meals.

Fifty other potentially infected ticks, not sacrificed for culture attempts, were allowed to refeed as newly emerged adults on two noninfected dogs. Following corticosteroid treatment 1 year after these ticks had fed, cultures of bone marrow from the dog exposed to ticks infected with the Oklahoma parasite were *Leishmania*-positive. At the same time, xenodiagnosis using laboratory bred ticks also demonstrated leishmanial infection. Other attempts to isolate or observe leishmaniae in this dog were negative. *Leishmania* was not detected in the other dog used in this uninterrupted tick feeding trial, nor in sylvatic mammals that occur near the kennels at Edmond and Stanley (J. Fox, personal communication). Serological evidence of leishmaniasis in coyotes *(Canis latrans)* in nearby Texas (42) may or may not have a bearing on transmission on Oklahoma or Kansas; it must first be demonstrated that these animals actually harbor *Leishmania*. The role of sand flies as vectors cannot be discounted, especially in view of the occurrence of *L. diabolica,* a mammal-feeder, in northern Texas (192).

In 1984, we had the opportunity to study the development of *Le. d. infantum* in experimentally infected females of this sand fly (unpublished data). We allowed 57 laboratory-reared *L. diabolica* to feed on the ears, back, and stomach of a dog that had previously lived in Sicily for 3 years, but which had returned to Florida with its owners. It was assumed to be infected with *Le. d. infantum* because of its history and clinical symptoms. Also, the dog had not lived in endemic areas other than Sicily. There were abundant amastigotes in the healthy-appearing skin and in the spleen and bone marrow.

Following the infecting blood meals, the sand flies were maintained at 27°C, 85–95% RH. An overall infection rate of 89.5% (51 of 57 flies) was subsequently recorded. We observed massive infections of the stomodeal valve, beginning at day 5 postfeed, but not in the cibarium or mouth parts at any time, up to 11 days after the infecting meal. Eleven of these flies, with infections ranging from 6 to 11 days, refed on noninfected hamsters and on one BALB/c mouse. Transmission was not demonstrated when these animals were examined 11–12 months later. Two other hamsters inoculated with amastigotes from bone marrow aspirates from the same dog were found heavily infected when they died 6–8 months after inoculation.

During the time observed (up to 11 days), the growth of this parasite in *L. diabolica* was essentially identical to that observed in *Phlebotomus ariasi,* a natural vector of *Le. d. infantum* in France (138), and in *L. longipalpis,* experimentally infected with the same parasite (68). According to these studies, promastigotes do not ordinarily invade the pharynx until 10 days of extrinsic growth and they do not appear in the mouth parts until after day 15. None of the *L. diabolica* in our study survived longer than 11 days, so it was impossible to determine if further development would have occurred. However, since there was no indication on of abnormal growth or parasite mortality in any of the infected flies, we regard *L. diabolica* as a potential vector of this disease.

Section Suprapylaria (*Leishmania hertigi* Complex)

The two named subspecies in this complex, *Le. hertigi hertigi* and *Le. h. deanei,* asymptomatically infect tree porcupines in the neotropics, but have not been isolated from human patients or phlebotomine sand flies (72). The nominate subspecies was described from Panamanian porcupines, *Coendou rothschildi,* in which infection rates as high as 88% have been recorded (49). *Leishmania h. deanei* parasitizes two other species of *Coendou* in Brazil (76).

Section Suprapylaria (*Leishmania mexicana* Complex)

The only proven vectors of the *Le. mexicana* parasites belong in the *flaviscutellata* complex of *Lutzomyia* (34, 95), but undoubtedly sand flies in other species groups will be incriminated in some foci. For example, no species of the *flaviscutellata* complex species, for example, occur in southcentral Texas or in the Dominican Republic, where autochthonous human cases of leishmaniasis caused by *Le. mexicana* or *Le. mexicana*-like parasites have been reported. This group of *Leishmania* is represented by one or more taxa in many parts of the lowland neotropics and in the southern Nearctic region of northern Mexico and Texas.

Leishmania Mexicana Mexicana

From intensive studies carried out in southern Mexico (10) and Belize (182) mostly before 1970, our knowledge of the epidemiology of leishmaniasis caused by this subspecies is relatively good. It is an enzootic disease, normally circulating among forest rodents, especially *Ototylomys phyllotis, Heteromys desmarestianus,* and *Nyctomys sumichrasti* (76). The incriminated vector is *L. o. olmeca,* distributed in lowland forests from Mexico to Honduras (34) and Costa Rica (J. Murillo, unpublished observations). In the neotropics the disease is known only in Mexico, Belize, Guatemala, and possibly Honduras (171). The risk of human infection is especially great during the rainy seasons, when adult populations of *L. o. olmeca* are high (182). Disney (73) found the first naturally infected female of this subspecies in Belize; others were subsequently discovered in that country (182) and in Mexico (10). The habits of this sand fly were reviewed by Williams (182) and Lainson and Shaw (76).

Recently, a wild-caught female of *L. ylephiletor* was found infected with *Le. m. mexicana* in Izabal Department, Guatemala (C. Porter, personal communication). The identification of the strain was confirmed by enzyme electrophoresis (R. Kreutzer, personal communication). The significance of this finding in a highly anthropophilic sand fly, unlike *L. o. olmeca,* can be determined only when further studies are undertaken.

Leishmania m. mexicana or a very similar parasite (70) also occurs in northern Mexico and southcentral Texas, where the climate and vegetation differ sharply from those of neotropical foci. There is now little doubt that the human cases of cutaneous and diffuse cutaneous leishmaniasis reported in Coahuila State, Mexico [three cases (128)] and in Texas [eight cases (44)] were autochthonous. In 1985, a boy from Uvalde, Texas, and a domestic cat from the same city but a different household were parasitologically confirmed as having cutaneous leishmaniasis (T. Gustafson, personal communication). Leishmaniae isolated from them are now being biochemically identified. Neither the boy nor the cat had traveled to leishmaniasis-endemic areas outside of Texas.

Lutzomyia diabolica, closely related to, but distinct from, *L. cruciata* (192), is the suspected vector in these foci because it is the only known anthropophilic sand fly present (192); its geographic distribution parallels that of known human infections (29, 44); and females are able to transmit the parasite following extrinsic development under experimental conditions (86). The latter evidence for vector incrimination is given less weight than the others because *L. anthophora* and *L. shannoni* females are also able to transmit experimentally one of the Texas-isolated parasites (86, 192). The former species is not anthropophilic (31) and *L. shannoni* does not exist in these endemic areas (192). No natural leishmanial infections have been found in Texas sand flies or in vertebrates other than man and the domestic cat.

It is significant that most patients in Texas probably contracted the disease in late summer or during autumn (44), when the proportion of parous, and thus infective, flies may be the highest (43). In Uvalde County, Texas, adults of *L. diabolica* have been found from May 17 to December 14 (86). Other information on the habits of this species (86) include the following: Females are agressive man-biters, attacking man mostly during the evening hours, but occasionally during the day as well. The species is exophilic and endophilic and is commonly found resting in houses and outbuildings. The natural resting sites and host preference range remain unknown. Females are anautogenous and gonotropically concordant. Nearly all eggs laid by late-season females (November–December) will not hatch until 90–270 days later. In contrast, those laid by females in June and July all hatch within 5–25 days following oviposition. Specimens may also overwinter as mature larvae.

Of 523 laboratory-bred *L. diabolica* females that fed on leishmaniasis-infected hamsters (MHOM/US/80/WR-411), 460 (87.9%) became infected (86). Suprapylarian development was confirmed in all females that died before refeeding. Short, active promastigotes were observed in the mouth parts beginning as early as day 4 postfeeding. Twenty-five females survived the first oviposition to refeed on 21 noninfected hamsters, five of which subsequently developed leishmanial lesions from 25 to 148 days after being bitten. It was shown that flies previously fed on infected hamsters from 4 to 7 days earlier transmitted this strain of *Leishmania* by bite to uninfected hamsters (86).

Leishmania Mexicana Amazonensis

From the extensive studies undertaken in Pará State, Brazil (76, 132, 155, 173, 175), much information has been obtained on the natural transmission cycle and vector relationships of *Le. m. amazonensis,* considered by some to be identical to *Le. m. pifanoi* (116). This parasite is transmitted to terrestrial rodents, especially *Proechimys guyannensis* and to a lesser degree to other mammals by *L. flaviscutellata,* a sand fly only slightly anthropophilic (76). For this reason, human disease due to this *Leishmania* is relatively rare, accounting for fewer than 3% of all leishmaniasis cases seen at one laboratory in Belém, Brazil from 1965 to 1982 (73). Recently near Manaus, Brazil, *L. olmeca nociva* females were found naturally infected with *Le. m. amazonensis* (J. Arias, personal communication). Both simple cutaneous and a suprisingly high proportion (~30%) of diffuse cutaneous leishmaniasis are caused by *Le. m. amazonensis* (73).

Up to 1982, 45 of 7498 *L. flaviscutellata* females examined had natural infections of this *Leishmania* (73). Differences in infection rates of these sand flies collected within 300 m apart in north Brazil suggested that "hot spots" of infections exist, probably because of the proximity of infected reservoir hosts (76). During these investigations, laboratory colonies of *L. flaviscutellata* were established (170) and experimentally infected females transmitted *Le. m. amazonensis* by bite to hamsters (175).

Elsewhere, *Le. m. amazonensis* has been isolated from spiny rats, *Proechimys,* and man in French Guiana (28a) and probably from a *L. flaviscutellata* female in southern Venezuela (125). The identity of the sand fly from Venezuela was never confirmed and it may have represented one of the *L. olmeca* taxa, an undescribed subspecies of which was recently discovered in Amazonas, Venezuela (D. Feliciangeli, personal communication).

This *Leishmania,* or a similar one, was found in field-collected females of *L. flaviscutellata* in Trinidad (165), but no natural infections in this species have been reported in Colombia, Surinam, French Guiana, Ecuador, and Peru, near the Bolivian border, where *L. flaviscutellata* is known to occur (106, 192). Recent characterization of *Le. m. amazonensis* from human patients in Colombia (R. Kreutzer, personal communications) indicates that this parasite may be widely distributed in the Amazon basin.

Leishmania Mexicana Aristedesi

This distinctive subspecies (24) is known only from Sasardi, San Blas Province, Panama, where Herrer *et al.* (54) identified infections in wild-caught rodents, especially *Oryzomys capito* (14 positive of 39 examined). Nearly all sand flies (99%) captured in rodent-baited Disney traps at this lowland site (22, 159) represented *L. o. bicolor,* a subspecies also known from Costa Rica, Colombia, Venezuela, and Ecuador (12, 190). This sand fly may also occur in some parts of Brazil, Peru, and possibly Bolivia, but more collections are needed for confirmation (D. Young, unpublished data). Christensen *et al.* (22) failed to find natural infections of *Le. m. aristedesi* in 449 female *L. o. bicolor* dissected and in over 1000 females pooled; nevertheless, they considered that this sand fly is the probable vector at Sasardi due to its feeding habits, abundance, and close taxonomic relationship to proven vectors in the *L. flaviscutellata* complex. It is not a common man-biter in Panama (24) and *Le. m. aristedesi* has not yet been isolated from humans there or elsewhere.

Leishmania Mexicana Enreittii

This subspecies parasitizes domestic guinea pigs, and *Lutzomyia monticola* has been experimentally infected (102); both the sand fly and the parasite occur at or near Curitiba, Paraná State, Brazil (73), but further studies are needed to incriminate this sand fly as a natural vector.

Leishmania Mexicana Garnhami

The taxonomic status of this parasite, isolated from human patients and from a *Didelphis* opossum (73, 153) in Trujillo State, Venezuela, remains questionable. It has been determined that strains from there are indistinguishable from *Le. m. amazonensis* by enzyme electrophoresis and other analyses (116). Moreover, the possibility exists that a *Le. braziliensis* subspecies may also occur in the endemic zone due to the slow *in vitro* growth of some stocks and because of observed peripylarian growth in some experimentally infected sand flies (4a, 119).

However, it seems clear that the suspected vector is an undescribed *Lutzomyia* species in the *verrucarum* species group and not *L. townsendi* as previously believed (D. Feliciangeli, J. Murillo, and D. Young, unpublished data). This sibling species of *L. townsendi* also occurs in Costa Rica, where it is also suspected of being a leishmaniasis vector (R. Zeledón, personal communication).

The altitudinal range of this anthropophilic species in the Andes of western Venezuela correlates well with that of human infections, mainly from 800 to 1800 m above sea level, where it is the most commonly encountered sand fly (114). Further information on its habits and relationship to *Leishmania* is given in other articles (4a, 17, 105, 119, 152). Naturally infected females have not been reported from western Venezuela, but experimentally infected females have successfully transmitted *Le. m. garnhami* to hamsters by bite (151a).

There is no basis for the suggestion (73) that a species in the *L. flaviscutellata* complex may be involved in the transmission of *Le. m. garnhami* or other *Leishmania* at these relatively high elevations.

Leishmania Mexicana Pifanoi

There is mounting evidence that this parasite may be consubspecific with *Le. m. amazonensis,* and, if confirmed, then the latter taxon will become a junior synonym (116). *Leishmania m. pifanoi* has been isolated only from human patients suffering from diffuse cutaneous leishmaniasis in Venezuela (72).

Leishmania Mexicana Venezuelensis

Since 1974, 73 human cases of leishmaniasis caused by this parasite have been reported in the Barquisimeto, Venezuela area (12). At one site, "Bosque de Macuto," six anthropophilic *Lutzomyia* species, including *L. olmeca bicolor,* have been captured (12). There is no other information on potential vectors.

Leishmania Mexicana Subspecies from Mato Grosso State, Brazil

This apparently distinct *Leishmania* (72) is known only from strains isolated from human patients (72) and there is no information on vectors.

Leishmania Mexicana Subspecies from Minas Gerais State, Brazil

This parasite has been isolated from man and dogs and, again, nothing is known about associated vectors (72).

Leishmania Species from the Dominican Republic

Bogaert-Díaz *et al.* (11) first reported autochthonous human cases of leishmaniasis from the Dominican Republic in 1975. Each of the three brothers had diffuse cutaneous leishmaniasis (DCL), a relatively rare condition observed in individuals with deficient cell-mediated immunity and, in the New World, associated with *Le. mexicana* parasites (178). Lainson (73) reported that one leishmanial strain from the Dominican Republic showed typical suprapylarian development in experimentally infected *L. longipalpis*. Further characterization indicated that the Dominican *Leishmania* is closely related to other *Le. mexicana* parasites, but differs in enzyme profiles and other characteristics (151).

Since 1975, 22 additional patients, all presenting DCL, have been parasitologically confirmed in the Dominican Republic (61). Sixteen of them probably, or certainly, contracted the disease in the eastern mountains (Cordillera Oriental) in the provinces of El Seibo and Altagracia (61). No simple ulcerated lesions have been observed.

In 1949, Fairchild and Trapido (35) collected two *Lutzomyia* spp. in Hispaniola, which they described and named *L. christophei (verrucarum* species group) and *L. cayennensis hispaniolae (cayennensis* species group). They remain the only extant sand flies known from Hispaniola.

In view of this unique and little-known focus of leishmaniasis, Johnson (61) conducted a sand fly survey in the Dominican Republic at various times from 1981 to 1983. The following comments are based on his observations, except where otherwise cited.

Of 17 case sites visited one or more times, Johnson noted that all were associated with shady coffee/cacoa groves, which simulated forests that once covered most of the island. Residents at some of these case sites are aware of biting flies ("erisos"), describing them as small and pale, with a hopping flight, and that begin to bite at dusk, inside and outside of houses. The probability that "erisos" were phlebotomines was confirmed when a DCL patient collected two female "erisos" biting him in September 1983. The flies were later identified as *L. christophei* by R. Johnson and D. Young. This observation remains the only one showing that *L. christophei* females bite humans under natural conditions.

Other observations from Johnson's study support the hypothesis that this species is the probable vector. In contrast to *L. c. hispaniolae,* which feeds on reptiles (mostly lizards), females of *L. christophei* are mammalophilic and were observed to feed on wild-caught rats *(Rattus rattus),* albino laboratory mice, and hamsters. Laboratory-reared females probed the skin of human volunteers, but did not take full blood meals.

Furthermore, one leishmanial stock (Isabel strain, WR 336) developed and multiplied well in experimentally infected females of *L. christophei,* showing typical suprapylarian growth. Eight of these 52 females, fed previously on infected BALB/c mice, took second blood meals from three noninfected mice 7–12 days after the infecting feeds. Four weeks later, one of the mice had a slight swelling of the hind foot where one of the *L. christophei* females had earlier fed. Xenodiagnosis using laboratory-bred females of *L. anthophora* demonstrated leishmanial infection. In addition, *in vitro* cultures of hindfoot aspirates from this mouse and from spleen tissue of the two others were also *Leishmania*-positive.

Johnson collected small numbers of *L. christophei* at or near seven human case sites from tree holes, on tree trunks, and in rock crevices. At no time during the study were they common; 23 wild-caught females were examined for leishmanial infections, but none was positive. No flies were captured in hamster-baited Disney traps placed close to known resting sites, nor were any collected from human bait. However, these collection techniques were not systematically used during the entire study period.

The natural hosts of *L. christophei* remain unstudied. Introduced murine rodents *(Rattus* and *Mus)* have largely replaced native rodents, which may have served as the original hosts of this *Leishmania*. The only two remaining mammals that are precinctive to Hispaniola are *Plagiodontia aedium,* a large, secretive rodent, and *Solenodon paradoxus,* an insectivore (187). Both species are uncommon and do not live in close proximity to man. Other potential reservoirs include the introduced mongoose, dogs, feral cats, or, less likely, man himself. Johnson, using the indirect fluorescent antibody test, found that 4 of 44 *Rattus rattus* captured near human case sites were seropositive for leishmanial antibodies, but no amastigotes were observed or cultured from any of them.

Concluding Remarks

An increased awareness of the public health importance of the leishmaniases during the past few years (178) has stimulated much research on nearly all aspects of the diseases. Knowledge of the vectors and their relationship with *Leishmania* has lagged behind that of mosquitoes and their associated pathogens, but some of the difficulties in studying these small insects have been overcome. Productive laboratory colonies of additional *Lutzomyia* vectors have been established recently. Forthcoming results of genetic and physiological studies of these colonized specimens should help determine the specific determinants that affect vector competency of sand flies—a salient, but little understood subject.

In general, nearly all topics on vectors covered or not covered in this chapter require considerably more attention. Long-term epidemiological studies are needed for a complete understanding of leishmanial transmission cycles and to monitor changes in disease prevalence brought about by changes in habitat. Such commitments are essential for the formulation of control measures, which, with few exceptions, have been completely unsuccessful in the New World (104).

Acknowledgments. We are deeply indebted to our colleagues, mentioned throughout the text, for generously providing unpublished information.

This review was supported by U.S. Army Medical Research and Development Command Contract DAMD-17-82-C-2223 and the National Institute of Allergy and Infectious Diseases, NIH, USPHS, Contract 5P01A120108-02. The views of the authors do not purport to reflect the position of the Department of the Army or the Department of Defense.

References

1. Abonnenc, E., and Léger, N., 1976, Sur une classification rationnelle des Diptères Phlebotomidae, *Cah. ORSTOM Ser. Entomol. Med. Parasitol.* **14**:69–78.
2. Abonnenc, E., and Léger, N., 1977, Rectificatif à la note: "Sur une classification rationnelle des Diptères Phlebotomidae," *Cah. ORSTOM Ser. Entomol. Med. Parasitol.* **14**:357.
3. Aguilar, C.M., Fernandez, E., de Fernandez, R., and Deane, L.M., 1984, Study of an outbreak of cutaneous leishmaniasis in Venezuela. The role of domestic animals, *Mem. Inst. Oswaldo Cruz* **79**:181–195.
4. Anderson, D.C., Buckner, R.G., Glenn, B.L., and MacVean, D.W., 1980, Endemic canine leishmaniasis, *Vet. Pathol.* **17**:94–96.

4a. Añez, N., Nieves, E., and Scorza, J.V., 1985, El "status" taxonomico de *Leishmania garnhami,* indicado por su patron de desarrollo en el vector., *Mem. Inst. Oswaldo Cruz* **80:**113–119.
5. Arias, J.R., and de Freitas, R.A., 1978, Sôbre as vetores da leishmaniose cutânea na Amazônia central do Brazil. 2. Incidencia de flagelados en flebótomos selváticos, *Acta Amazonica* **8:**387–396.
6. Arias, J.R., and Naiff, R.D., 1981. The principal reservoir host of cutaneous leishmaniasis in the urban areas of Manaus, Central Amazon of Brazil, *Mem. Inst. Oswaldo Cruz* **76:**279–286.
7. Arias, J.R., Naiff, R.D., Miles, M.A., and de Souza, A.A., 1981, The opossum, *Didelphis marsupialis* (Marsupialia: Didelphidae), as a reservoir host of *Leishmania braziliensis guyanensis* in the Amazon basin of Brazil, *Trans. R. Soc. Trop. Med. Hyg.* **75:**537–541.
8. Arias, J.R., Miles, M.A., Naiff, R.D., Póvoa, M.M., de Freitas, R.A., Biancardi, C.B., and Castellon, E.G., 1985, Flagellate infections of Brazilian sandflies (Diptera: Psychodidae): Isolation *in vitro* and biochemical identification of *Endotrypanum* and *Leishmania, Am. J. Trop. Med. Hyg.* **34:**1098–1108.
9. Beach, R.F., Kiilu, G., and Leeuwenburg, J., 1985, Modification of sand fly behavior by *Leishmania* leads to increased parasite transmission, *Am. J. Trop. Med. Hyg.* **34:**278–282.
10. Biagi, F., Biagi, A.M., and Beltran, F., 1965, *Phlebotomus flaviscutellatus* transmisor natural de *Leishmania mexicana, Prensa Med. Mex.* **30:**267–272.
11. Bogaert-Díaz, H., Rojas, R.F., de León, A., de Martinez, D., and de Quiñones, M., 1974, Leishmaniasis tegumentaria Americana. Reporte de los primeros tres casos descubiertos en R.D. forma anergica en tres hermanos, *Rev. Domin. Dermatol.* **9:**19–31.
12. Bonfante-Garrido, R., 1984, Endemic cutaneous leishmaniasis in Barquisimeto, Venezuela, *Trans. R. Soc. Trop. Med. Hyg.* **78:**849–850.
13. Bray, R.S., 1981, Travellers in Peril: The promastigote and the sporozoite, in: E.U. Canning (ed.), *Parasitological Topics, A Presentation Volume to P.C. Garnham F.R.S. on the Occasion of His 80th Birthday,* Allen Press, Lawrence, Kansas, pp. 48–53.
14. Buescher, M.D., Rutledge, L.C., Roberts, J., and Nelson, J.H., 1984, Observations on multiple feedings of *Lutzomyia longipalpis* in the laboratory (Diptera: Psychodidae), *Mosq. News* **44:**76–77.
15. Burkot, T.R., Williams, J.L., and Schneider, I., 1984, Identification of *Plasmodium falciparum*-infected mosquitoes by a double antibody enzyme-linked immunosorbent assay, *Am. J. Trop. Med. Hyg.* **33:**783–788.
16. Butler, J.F., Kloft, W.J., DuBose, L.A., and Kloft, E.S., 1977, Recontamination of food after feeding a ^{32}P food source to biting Muscidae, *J. Med. Entomol.* **13:**567–571.
16a. Caillard, T., Tibayrenc, M., Le Pont, F., Dujardin, J.P., Desjeux, P., and Ayala, F.J., 1986, Diagnosis by isozyme methods of two cryptic species, *Psychodopygus carrerai* and *P. yucumensis* (Diptera: Psychodidae), *J. Med. Entomol.* **23:**489–492.
17. Carnevali, M., and Scorza, J.V., 1982, Factores dermicos que condicionan la infección de *Lutzomyia townsendi* (Ortiz, 1959) por *Leishmania* spp. de Venezuela, *Mem. Inst. Oswaldo Cruz* **77:**353–365.

18. Chaniotis, B.N., Correa, M.A., Tesh, R.B., and Johnson, K.M., 1974, Horizontal and vertical movement of phlebotomine sandflies in a Panamanian rain forest, *J. Med. Entomol.* **11**:369–375.
19. Christensen, H.A., and Herrer, A., 1972, Detection of *Leishmania braziliensis* by xenodiagnosis, *Trans. R. Soc. Trop. Med. Hyg.* **66**:798–799.
20. Christensen, H.A., and Herrer, A., 1979, Susceptibility of sand flies (Diptera; Psychodidae) to Trypanosomatidae from two-toed sloths (Edentata: Bradypodidae), *J. Med. Entomol.* **16**:424–427.
21. Christensen, H.A., and Herrer, A., 1980, Panamanian *Lutzomyia* (Diptera: Psychodidae) host attraction profiles, *J. Med. Entomol.* **17**:522–528.
22. Christensen, H.A., Herrer, A., and Telford, S.R., 1972, Enzootic cutaneous leishmaniasis in eastern Panama. II. Entomological investigations, *Ann. Trop. Med. Parasitol.* **66**:55–66.
23. Christensen, H.A., Arias, J.R., de Vasques, A.M., and de Freitas, R., 1982, Hosts of sand fly vectors of *Leishmania braziliensis guyanensis* in the central Amazon of Brazil, *Am. J. Trop. Med. Hyg.* **31**:239–242.
24. Christensen, H.A., Fairchild, G.B., Herrer, A., Johnson, C.M., Young, D.G., and de Vasquez, A.M., 1983, The ecology of cutaneous leishmaniasis in the Republic of Panama, *J. Med. Entomol.* **20**:463–484.
25. Coehlo, M.V., and Falcão, A.R., 1962, Transmissão experimental de *Leishmania braziliensis*. II. Transmissão de amostra mexicana por picada de *Phlebotomus longipalpis* e de *Phlebotomus renei, Rev. Inst. Med. Trop. S. Paulo* **4**:220–224.
26. Courmes, E., Escudie, A., Fauran, P., and Monnerville, A., 1966, Premier cas autochtone de leishmaniose viscérale humaine a la Guadeloupe, *Bull. Soc. Pathol. Exot.* **59**:217–266.
27. Dallai, R., Baccetti, B., Mazzini, M., and Sabatinelli, G., 1984, The spermatozoon of three species of *Phlebotomus* (Phlebotominae) and the acrosomal evolution in nematoceran Dipterans, *Int. Insect Morphol. Embryol.* **13**:1–10.
28. Decker-Jackson, J.E., and Tang, D.B., 1982, Identification of *Leishmania* spp. by radiorespirometry II. A statistical method of data analysis to evaluate the reproducibility and sensitivity of the technique, in: M.L. Chance and B.C. Walton (eds.), *Biochemical Characterization of Leishmania,* UNDP/World Bank/WHO, pp. 205–245.
28a. Dedet, J.P., Pradinaud, R., Desjeux, P., Jacquet-Viallet, P., Girardeau, I., Esterre, P., and Gotz, W., 1985, Deux primiers cas de leishmaniose cutanée à *Leishmania mexicana amazonensis* en Guyane Française, *Bull. Soc. Path. Exot.,* **78**:64–70.
29. Díaz-Nájera, A., 1971, Precencia de *Lutzomyia (Lutzomyia) diabolica* (Hall, 1936) en Muzquiz, Coahuila, Mexico (Diptera, Psychodidae), *Rev. Invest. Salud. Publica* **31**:62–66.
30. Dollahon, N.R., and Janovy, J., 1974, Experimental infection of New World lizards with Old World *Leishmania* species, *Exp. Parasitol.* **36**:253–260.
31. Endris, R.G., Perkins, P.V., Young, D.G., and Johnson, R.N., 1982, Techniques for laboratory rearing of sand flies (Diptera: Psychodidae), *Mosq. News* **42**:400–407.
32. Endris, R.G., Young, D.G., and Butler, J.F., 1984, The laboratory biology of the sand fly *Lutzomyia anthophora* (Diptera: Psychodidae), *J. Med. Entomol.* **21**:656–664.

33. Evans, D.A., Lanham, S.M., Baldwin, C.I., and Peters, W., 1984, The isolation and isoenzyme characterization of the *Leishmania braziliensis* subsp. from patients with cutaneous leishmaniasis acquired in Belize, *Trans. R. Soc. Trop. Med. Hyg.* **78**:35–42.
34. Fairchild, G.B., and Theodor, O., 1971, On *Lutzomyia flaviscutellata* (Mangabeira) and *L. olmeca* (Vargas and Díaz-Nájera) (Diptera: Psychodidae), *J. Med. Entomol.* **8**:153–159.
35. Fairchild, G.B., and Trapido, H., 1950, The West Indian species of *Phlebotomus* (Diptera, Psychodidae), *Ann. Entomol. Soc. Am.* **43**:405–417.
36. Forattini, O.P., 1971, Sôbre a classificação da subfamilia Phlebotominae nas Américas (Diptera, Psychodidae), *Pap. Avulsos Zool. (Sao Paulo)* **24**:93–111.
37. Forattini, O.P., 1973, *Entomologia Médica. IV. Psychodidae, Phlebotominae, Leishmanioses, Bartonelose,* Edgard Blucher, São Paulo. 658 pp.
38. Forattini, O.P., Pattoli, D.B.G., Rabello, E.X., and Ferreira, O.A., 1972, Infecção natural de phlebotomíneos em foco enzoótico de leishmaniose tegumentar no estado de São Paulo, Brasil, *Rev. Saude Pública* **6**:431–433.
39. Franke, E.D., Rowton, E.O., McGreevy, P.B., and Perkins, P.V., 1984, Detection and enumeration of *Leishmania* amastigotes and promastigotes in sandflies using agar plates, *Abstr. Annu. Meet. Am. Soc. Trop. Med. Hyg., 33rd, Baltimore,* p. 179.
40. Frankenburg, S., Londner, M.V., Schlein, Y., and Schnur, L.F., 1985, The development of a solid phase radioimmunoassay for the detection of leishmanial parasites in the sand fly, *Am. J. Trop. Med. Hyg.* **34**:266–269.
41. Gentile, B., LePont, F., Pajot, F.-X., and Besnard, R., 1981, Dermal leishmaniasis in French Guiana: The sloth *(Cholopeus didactylus)* as a reservoir host, *Trans. R. Soc. Trop. Med. Hyg.* **75**:612–613.
42. Grogyl, M., Kuhn, R.E., Davis, D.S., and Green, G.E., 1984, Antibodies to *Trypanosoma cruzi* in coyotes in Texas, *J. Parasitol.* **70**:189–191.
43. Guilvard, E., Wilkes, T.J., Killick-Kendrick, R., and Rioux, J.A., 1981, Écologie des leishmanioses dans le sud de la France. 16. Déroulement des cycles gonotrophiques chez *Phlebotomus ariasi* Tonnoir, 1921 et *Phlebotomus mascittii* Grassi, 1908 en Cévennes. Corollaire épidemiologique, *Ann. Parasitol. Hum. Comp.* **55**:659–664.
44. Gustafson, T.L., Reed, C.M., McGreevy, P.B., Pappas, M.G., Fox, J.C., and Lawyer, P.G., 1985, Human cutaneous leishmaniasis acquired in Texas, *Am. J. Trop. Med. Hyg.* **34**:58–63.
45. Hanson, W.J., 1961, The breeding places of *Phlebotomus* in Panama (Diptera, Psychodidae), *Ann. Entomol. Soc. Am.* **54**:317–322.
46. Hanson, W.J., 1968, The immature stages of the subfamily Phlebotominae in Panama (Diptera, Psychodidae), Ph.D. thesis, University of Kansas, Lawrence, Kansas. 104 pp.
47. Hashiguchi, Y., Landires, E.A.G., de Coronel, V.V., Mimori, T., and Kawabata, M., 1985. Natural infections with promastigotes in man-biting species of sandflies in leishmaniasis endemic areas of Ecuador, *Am. J. Trop. Med. Hyg.* **34**:440–446.
48. Herrer, A., 1951, Estudios sobre leishmaniasis tegumentaria en el Peru. II. Infeccion experimental de zorros con cultivos de leishmanias aisladas de casos de uta, *Rev. Med. Exp. (Lima)* **8**:29–33.

49. Herrer, A., 1971, *Leishmania hertigi* sp. n., from the tropical porcupine, *Coendou rothschildi* Thomas, *J. Parasitol.* **57:**626–629.
50. Herrer, A., 1982a, Empleo del hamster dorado como animal centinela en las localidades donde es endémica la uta (Leishmaniasis tegumentaria), *Rev. Inst. Med. Trop. S. Paulo* **24:**162–167.
51. Herrer, A., 1982b, *Lutzomyia peruensis* (Shannon, 1929) possible vector natural de la uta (Leishmaniasis tegumentaria), *Rev. Inst. Med. Trop. S. Paulo* **24:**168–172.
52. Herrer, A., and Christensen, H.A., 1975a, Infrequency of gross skin lesions among Panamanian forest mammals with cutaneous leishmaniasis, *Parasitology,* **71:**87–92.
53. Herrer, A., and Christensen, H.A., 1975b, Implication of *Phlebotomus* sand flies as vectors of bartonellosis and leishmaniasis as early as 1764, *Science* **190:**154–155.
54. Herrer, A., Telford, S.R., and Christensen, H.A., 1971, Enzootic cutaneous leishmaniasis in eastern Panama I: Investigation of the infection among forest mammals, *Ann. Trop. Med. Parasitol.* **65:**349–358.
55. Herrer, A., Christensen, H.A., and Beumer, R.J., 1973, Reservoir hosts of cutaneous leishmaniasis among Panamanian forest mammals, *Am. J. Trop. Med. Hyg.* **22:**585–591.
56. Hertig, M., 1942, *Phlebotomus* and carrion's disease, *Am. J. Trop. Med.* **22**(Suppl.):1–81.
56a. Hudson, J.E., and Young, D.G., 1985, New records of phlebotomines, leishmaniasis, and mosquitoes from suriname, *Trans. R. Soc. Trop. Med. Hyg.* **79:**418.
57. Jackson, P.D., Wohlhieter, J.A., and Hockmeyer, W.T., 1982, *Leishmania* characterization by restriction endonuclease digestion of kinetoplastic DNA, *Abstr. Intl. Congr. Parasitol. 5th, Toronto,* p. 342.
58. Jobling, B., 1976, On the fascicle of blood-sucking Diptera in addition to a description of the maxillary glands in *Phlebotomus papatasi* together with the musculature of the labium and pulsatory organ of both the latter species and some other Diptera, *J. Nat. Hist.* **10:**457–461.
59. Johnson, P.T., 1961, Autogeny in Panamanian *Phlebotomus* sandflies (Diptera: Psychodidae), *Ann. Entomol. Soc. Am.* **54:**116–118.
60. Johnson, P.T., and Hertig, M., 1961, The rearing of *Phlebotomus* sandflies (Diptera: Psychodidae). II. Development and behavior of Panamanian sandflies in laboratory culture, *Ann. Entomol. Soc. Am.* **54:**764–776.
61. Johnson, R.N., 1984, Phlebotomine sand flies (Diptera: Psychodidae) and diffuse cutaneous leishmaniasis in the Dominican Republic, Ph.D. thesis, University of Florida, Gainesville, Florida, 126 pp.
62. Kettle, D.S., 1985, *Medical and Veterinary Entomology,* Wiley, New York. 658 pp.
63. Killick-Kendrick, R., 1978, Recent advances and problems in the biology of phlebotomine sandflies, *Acta Trop.* **35:**297–313.
64. Killick-Kendrick, R., 1979, Biology of *Leishmania* in phlebotomine sandflies, in: W.H.R. Lumsden and D.A. Evans (eds.), *Biology of the Kinetoplastida,* Vol. 2, Academic Press, New York, pp. 306–460.
65. Killick-Kendrick, R., and Molyneux, D.H., 1981, Transmission of leishmaniasis by the bite of phlebotomine sandflies: Possible mechanisms, *Trans. R. Soc. Trop. Med. Hyg.* **75:**152–154.

66. Killick-Kendrick, R., and Ward, R.D., 1981, Ecology of *Leishmania,* Workshop No. 11, *Parasitology* **82:**143–152.
67. Killick-Kendrick, R., Leaney, A.J., and Ready, P.D., 1977, The establishment, maintenance and productivity of a laboratory colony of *Lutzomyia longipalpis* (Diptera: Psychodidae), *J. Med. Entomol.* **13:**429–440.
68. Killick-Kendrick, R., Molyneux, D.H., Rioux, J.A., and Leaney, A.J., 1980, Possible origins of *Leishmania chagasi, Ann. Trop. Med. Parasitol.* **74:**563–565.
69. Kimsey, R.B., and Kimsey, P.B., 1984, Identification of arthropod blood meals using rubidium as a marker: A preliminary study, *J. Med. Entomol.* **21:**714–719.
69a. Kreutzer, R.D., Modi, G.B., Tesh, R.B., and Young, D.G., 1987, Brain cell karyotypes of six species of New and Old World sand flies (Diptera: Psychodidae), *J. Med. Entomol.* **24:**(in press).
70. Kreutzer, R.D., Semko, M.E., Hendricks, L.D., and Wright, N., 1983, Identification of *Leishmania* spp. by multiple isozyme analysis, *Am. J. Trop. Med. Hyg.* **32:**703–715.
71. Lainson, R., 1982a, Leishmanial parasites of mammals in relation to human disease, in: M.A. Edwards and U. McDonnell (eds.), *Animal Disease in Relation to Animal Conservation,* Vol. 50, Symposia, Zoological Society of London, pp. 137–179.
72. Lainson, R., 1982b, *Leishmaniasis,* in: J.H. Steele (ed.), *CRC Handbook Series* in *Zoonoses, Series C: Parasitic Zoonoses,* Vol. 1, CRC Press, Boca Raton, Florida, pp. 41–103.
73. Lainson, R., 1983, The American leishmaniases: Some observations on their ecology and epidemiology, *Trans. R. Soc. Trop. Med. Hyg.* **77:**569–596.
74. Lainson, R., and Shaw, J.J., 1972a, Leishmaniasis of the New World: Taxonomic problems, *Br. Med. Bull.* **28:**44–48.
75. Lainson, R., and Shaw, J.J., 1972b, Taxonomy of the New World *Leishmania* species, *Trans. R. Soc. Trop. Med. Hyg.* **66:**943–944.
76. Lainson, R., and Shaw, J.J., 1979, The role of animals in the epidemiology of South American Leishmaniasis, in: W.H.R. Lumsden and D.A. Evans (eds.), *Biology of the Kinetoplastida,* Vol. 2, Academic Press, New York, pp. 1–116.
77. Lainson, R., and Southgate, B.A., 1965, Mechanical transmission of *Leishmania mexicana* by *Stomoxys calcitrans, Trans. R. Soc. Trop. Med. Hyg.* **59:**716.
78. Lainson, R., Shaw, J.J., Ward, R.D., and Fraiha, H. 1973. Leishmaniasis in Brazil. IX. Considerations of the *Leishmania braziliensis* complex: Importance of the sandflies of the genus *Psychodopygus* (Mangabeira) in the transmission of *L. braziliensis braziliensis* in North Brazil, *Trans. R. Soc. Trop. Med. Hyg.* **67:**184–196.
79. Lainson, R., Ward, R.D., and Shaw, J.J., 1977, Experimental transmission of *Leishmania chagasi,* causative agent of neotropical visceral leishmaniasis by the sand fly, *Lutzomyia longipalpis, Nature* **266:**628–630.
80. Lainson, R., Ready, P.D., and Shaw, J.J., 1979, *Leishmania* in phlebotomid sandflies. VII. On the taxonomic status of *Leishmania peruviana,* causative agent of Peruvian "uta", as indicated by its development in the sandfly, *Lutzomyia longipalpis, Proc. R. Soc. Lond. B* **206:**307–318.

81. Lainson, R., Shaw, J.J., Ward, R.D., Ready, P.D., and Naiff, R.D., 1979, Leishmaniasis in Brazil. XIII. Isolation of *Leishmania* from armadillos *(Dasypus novemcinctus)* and observations on the epidemiology of cutaneous leishmaniasis in North Pará State, *Trans. R. Soc. Trop. Med. Hyg.* **73**:239–242.
82. Lainson, R., Miles, M.A., and Shaw, J.J., 1981a, On the identification of viscerotropic leishmanias, *Ann. Trop. Med. Parasitol.* **75**:251–253.
83. Lainson, R., Shaw, J.J., Ready, P.D., Miles, M.A., and Póvoa, M., 1981b, Leishmaniasis in Brazil. XVI. Isolation and identification of *Leishmania* species from sandflies, wild mammals and man in North Pará State, with particular reference to *L. braziliensis guyanensis* causative agent of 'pian-bois'. *Trans. R. Soc. Trop. Med. Hyg.* **75**:530–536.
84. Lainson, R., Shaw, J.J., Ryan, L., Ribeiro, R.S.M., and Silveira, F.T. 1985, Leishmaniasis in Brazil. XXI. Visceral leishmaniasis in the Amazon region and further observations on the role of *Lutzomyia longipalpis* (Lutz and Neiva, 1912) as the vector, *Trans. R. Soc. Trop. Med. Hyg.* **79**:223–226.
84a. Lane, R.P., and Ready, P.D., 1985, Multivariate discrimination between *Lutzomyia wellcomei,* a vector of mucocutaneous leishmaniasis, and *Lu. complexus* (Diptera: Phlebotominae), *Ann. Trop. Med. Parasitol.,* **79**:469–472.
85. Lane, R., Phillips, A., Molyneux, D.H., Procter, G., and Ward, R.D., 1985, Chemical analysis of the abdominal glands of two forms of *Lutzomyia longipalpis:* Site of a possible sex pheromone?, *Ann. Trop. Med. Parasitol.* **79**:225–229.
86. Lawyer, P.G., 1984, Biology and colonization of the sand fly *Lutzomyia diabolica* (Hall) (Diptera: Psychodidae) with notes on its potential relationship to human cutaneous leishmaniasis in Texas, USA, Ph.D. thesis, University of Florida, Gainesville, Florida. 244 pp.
87. Léger, M., 1918, Infection sanguine par *Leptomonas* chez un saurien, *C. R. Soc. Biol. Filiales* **81**:772–774.
88. Lent, H., and Wygodzinsky, P. 1979. Revision of the Triatominae (Hemiptera, Reduviidae), and their significance as vectors of chagas disease, *Bull. Am. Mus. Nat. Hist.* **163**:125–520.
89. LePont, F., 1982, La leishmaniose en Guyane Française. 2. Fluctuations saisonnieres d'abondance et du taux d'infection naturelle de *Lutzomyia (Nyssomyia) umbratilis* Ward and Fraiha, 1977, *Cah. ORSTOM Ser. Entomol. Med. Parasitol.* **20**:269–277.
90. LePont, F., and Desjeux, P., 1985, Leishmaniasis in Bolivia, I. *Lutzomyia longipalpis* (Lutz and Neiva, 1912) as the vector of visceral leishmaniasis in Los Yungas, *Trans. R. Soc. Trop. Med. Hyg.* **79**:227–231.
91. LePont, F., and Pajot, F.-X., 1980, La leishmaniose en Guyane Française. I. Étude de l'ecologie et du taux d'infection naturelle du vecteur *Lutzomyia (Nyssomyia) umbratilis* Ward et Fraiha, 1977 en saison sèche. Considérations épidémiologiques, *Cah. ORSTOM Ser. Entomol. Med. Parasitol.* **28**:359–382.
92. LePont, F., and Pajot, F.-X. 1981. La leishmaniose en Guyane Française. 2. Modalités de la transmission dans un village forestier: Cacao, *Cah. ORSTOM Ser. Entomol. Med. Parasitol.* **19**:223–231.

93. Lewis, D.J., 1965, Internal structural features of some Central American phlebotomine sandflies, *Ann. Trop. Med. Parasitol.* **59:**375–385.
94. Lewis, D.J., 1974, The biology of Phlebotomidae in relation to leishmaniasis, *Annu. Rev. Entomol.* **19:**363–384.
95. Lewis, D.J., 1975a, The *Lutzomyia flaviscutellata* complex (Diptera: Psychodidae), *J. Med. Entomol.* **12:**363–368.
96. Lewis, D.J., 1975b, Functional morphology of the mouth parts in New World phlebotomine sandflies (Diptera: Psychodidae), *Trans. R. Entomol. Soc. Lond.* **126:**497–532.
97. Lewis, D.J., 1984, Trophic sensilla of phlebotomine sandflies, *Trans. R. Soc. Trop. Med. Hyg.* **78:**416.
98. Lewis, D.J., and Domoney, C.R., 1966, Sugar meals in Phlebotominae and Simuliidae, *Proc. R. Entomol. Soc. Lond.* **41:**175–179.
99. Lewis, D.J., Lainson, R., and Shaw, J.J., 1970, Determination of parous rates in phlebotomine sandflies with special reference to Amazonian species, *Bull. Entomol. Res.* **60:**209–219.
100. Lewis, D.J., Young, D.G., Fairchild, G.B., and Minter, D.M., 1977, Proposals for a stable classification of the phlebotomine sandflies (Diptera: Psychodidae), *Syst. Entomol.* **2:**319–332.
101. Llanos, B.Z., 1981, Los flebotomos del Peru y su distribution geografica (Diptera, Psychodidae, Phlebotominae), *Rev. Peru. Entomol.* **24:**183–184.
102. Luz, E., Giovannoni, M., and Borba, A.M., 1967, Infecção de *Lutzomyia monticola* por *Leishmania enriettii, An. Fac. Med. Univ. Fed. Parana* **9/10:**121–128.
103. Magnarelli, L.A., Modi, G.B., and Tesh, R.B., 1984, Follicular development and parity in phlebotomine sand flies (Diptera: Psychodidae), *J. Med. Entomol.* **21:**681–689.
104. Marinkelle, C.J., 1980, The control of leishmaniases, *Bull. WHO* **58:**807–818.
105. Marquez, M., and Scorza, J.V., 1982, Criterios de nuliparidad y de paridad en *Lutzomyia townsendi* (Ortiz, 1959) del accidente de Venezuela, *Mem. Inst. Oswaldo Cruz* **77:**229–246.
106. Martins, A.V., Williams, P., and Falcão, A.L., 1978, *American Sand Flies (Diptera: Psychodidae, Phlebotominae), Academia Brasileira de Ciencias,* Rio de Janeiro, Brazil. 195 pp.
107. McKenzie, K.K., 1984, A study of the transmission of canine leishmaniasis by the tick, *Rhipicephalus sanguineus* (Latreille) and an ultrastructural comparison of the promastigotes, Ph.D. thesis, Oklahoma State University, Stillwater, Oklahoma. 146 pp.
108. McMahon-Pratt, D., Bennett, E., and David, J.R., 1982, Monoclonal antibodies that distinguish subspecies of *Leishmania braziliensis, J. Immunol.* **129:**926–927.
109. McMahon-Pratt, D., Modi, G.B., and Tesh, R.B., 1983, Detection of promastigote stage-specific antigens on *Leishmania mexicana amazonensis* developing in the midgut of *Lutzomyia longipalpis, Am. J. Trop. Med. Hyg.* **32:**1268–1271.
110. Mendoza, J.L., Gonzalez, O.F., Rodriguez, M.C., Navarro, A., and Negrin, E.M., 1983, Estudio de la actividad hematofagica y el tiempo de ingesta de

Lutzomyia orestes (Diptera, Psychodidae) Informe preliminar, *Rev. Cubana Med. Trop.* **35:**357–362.
111. Miles, C.T., Foster, W.A., and Christensen, H.A., 1976, Mating aggregations of male *Lutzomyia* sandflies at human hosts in Panama, *Trans. R. Soc. Trop. Med. Hyg.* **70:**531–532.
112. Minter, D.M., and Goedbloed, E., 1971, The preservation in liquid nitrogen of tsetse flies and phlebotomine sandflies naturally infected with trypanosomatid flagellates, *Trans. R. Soc. Trop. Med. Hyg.* **65:**175–181.
113. Modi, G.B., and Tesh, R.B., 1983, A simple technique for mass rearing *Lutzomyia longipalpis* and *Phlebotomus papatasi* (Diptera: Psychodidae) in the laboratory, *J. Med. Entomol.* **20:**568–569.
114. Mogollon, J., Manzanilla, P., and Scorza, J.V., 1977, Distribucion altitudinal de nueve especies de *Lutzomyia* (Diptera: Psychodidae) en el Estado Trujillo, Venezuela, *Bol. Dir. Malariol. Saneam. Amb.* **17:**206–224.
115. Molyneux, D.H., 1983, Host–parasite relationships of trypanosomatidae in vectors, in: K.F. Harris (ed.), *Current Topics in Vector Research,* Vol. 1, Praeger, New York, pp. 117–148.
116. Momen, H., and Grimaldi, G., 1984, On the identity of *Leishmania pifanoi* and *L. mexicana garnhami, Trans. R. Soc. Trop. Med. Hyg.* **78:**701–702.
117. Morales, A., Corredor, A., Cacares, E., Ibagos, E., and de Rodriguez, C.I., 1981, Aislamiento de tres cepas de *Leishmania* a partir de *Lutzomyia trapidoi* in Colombia, *Biomedica (Bogota)* **1:**198–207.
118. Morales, A., de Carrasquilla, C.F., and de Rodriguez, C.I., 1984, Establecimiento de una colonia de *Lutzomyia walkeri* (Newstead, 1914) (Diptera: Phlebotominae), *Biomedica (Bogota)* **4:**37–41.
119. Moreno, E., and Scorza, J.V., 1981, Comportamiento *in vivo* e *in vitro* de seite aisladas de *Leishmania garnhami* del accidente de Venezuela, *Bol. Dir. Malariol. Saneam. Amb.* **21:**179–191.
120. Navin, T.R., Sierra, M., Custodio, R., Steurer, F., Porter, C.H., and Ruebush, T.K., 1985, Epidemiologic study of visceral leishmaniasis in Honduras, 1975–1983, *Am. J. Trop. Med. Hyg.* **34:**1069–1075.
121. Pajot, F.-X., LePont, F., Gentile, B., and Besnard, R., 1982, Epidemiology of leishmaniasis in French Guiana, *Trans. R. Soc. Trop. Med. Hyg.* **76:**112–113.
122. Perkins, P.V., 1982, The identification and distribution of phlebotomine sand flies in the United States with notes on the biology of two species from Florida (Diptera: Psychodidae), Ph.D. thesis, University of Florida, Gainesville, Florida. 196 pp.
123. Peterson, J.L., 1982, Preliminary survey of isozyme variation in anthropophilic Panamanian *Lutzomyia* species, in: M.L. Chance and B.C. Walton (eds.), *Biochemical Characterization of Leishmania,* UNDP/World Bank/WHO, pp. 104–114.
124. Pifano, F.C., Ortiz, I., and Alvarez, A., 1960, La ecologia, en condiciones naturales y de laboratorio, de algunas especies de *Phlebotomus* de la region de Guatopo, Estado Miranda, Venezuela, *Arch. Venez. Med. Trop. Parasitol. Med.* **3:**63–71.
125. Pifano, F.C., Morrell, J.R., and Alvarez, A., 1973, Comprobacion de una cepa de dermotropa en *Phlebotomus flaviscutellata* Mangabeira, 1942 de

Sierra Parima, Territorio Federal Amazonas, Venezuela, *Arch. Venez. Med. Trop. Parasitol. Med.* **5**:145–167.

126. Porter, C.H., and De Foliart, G.R., 1981, The man-biting activity of Phlebotomine sand flies (Diptera: Psychodidae) in a tropical wet forest environment in Colombia, *Arq. Zool. S. Paulo* **30**:81–158.
127. Quate, L.W., 1963, Fossil Psychodidae in Mexican amber, part 2 (Diptera: Insecta), *J. Paleontol.* **37**:110–118.
128. Ramos-Aguirre, C., 1970, Leishmaniasis en la región carbonifera de Coahuila. Reporte de dos casas de la forma anergica difusa, *Dermatol. Rev. Mex.* **14**:39–45.

128a. Rangel, E.F., Ryan, L., Lainson, R., and Shaw, J.J., 1985, Observations on the sandfly (Diptera: Psychodidae), fauna of Além Paraíba, State of Minas Gerais, Brazil; and The isolation of a parasite of the *Leishmania braziliensis* complex from *Psychodopygus hirsuta hirsuta, Mem. Inst. Oswaldo* Cruz. **80**:373–374.

129. Rangel, E.F., de Souza, N.A., Wermellnger, E.D., and Barbosa, A.F., 1984, Infecção natural de *Lutzomyia intermedia* Lutz and Neiva, 1912, em àrea endémica do leishmaniose tegumentar no Estado do Rio de Janeiro, *Mem. Inst. Oswaldo Cruz* **79**:395–396.
130. Ready, P.D., 1978, The feeding habits of laboratory-bred *Lutzomyia longipalpis* (Diptera: Psychodidae), *J. Med. Entomol.* **14**:545–552.
131. Ready, P.D., and da Silva, R.M.R., 1984, An alloenzymic comparison of *Psychodopygus wellcomei*—an incriminated vector of *Leishmania braziliensis* in Pará State, Brazil—and the sympatric morphospecies *Ps. complexus* (Diptera, Psychodidae), *Cah. ORSTOM Entomol. Med. Parasitol.* **22**:3–8.
132. Ready, P.D., Lainson, R., and Shaw, J.J., 1983a, Leishmaniasis in Brazil. XX. Prevalence of "enzootic rodent leishmaniasis" *(Leishmania mexicana amazonensis),* and apparent absence of "pian bois" *(Le. braziliensis guyanensis),* in plantations of introduced tree species and in other non-climax forests in eastern Amazonia, *Trans. R. Soc. Trop. Med. Hyg.* **77**:775–785.
133. Ready, P.D., Ribeiro, A.L., Lainson, R., de Alencar, J.E., and Shaw, J.J., 1983b, Presence of *Psychodopygus wellcomei* (Diptera: Psychodidae), a proven vector of *Leishmania braziliensis braziliensis,* in Ceará State, *Mem. Inst. Oswaldo Cruz* **78**:235–236.
134. Ready, P.D., Lainson, R., and Shaw, J.J., 1984a, Habitat and seasonality of *Psychodopygus wellcomei* help incriminate it as a vector of *Leishmania braziliensis* in Amazônia and northeast Brazil, *Trans. R. Soc. Trop. Med. Hyg.* **78**:543–544.

134a. Ready, P.D., Lainson, R., Shaw, J.J., and Ward, R.D., 1986, The ecology of *Lutzomyia umbratilis* Ward & Fraiha (Diptera: Psychodidae), the major vector to man of *Leishmania braziliensis guyanensis* in north-eastern Amazonian Brazil, *Bull. Entomol. Res.* **76**:21–40.

135. Ready, P.D., Lainson, R., Wilkes, T.J., and Killick-Kendrick, R. 1984b, On the accuracy of age-grading neotropical phlebotomines by counting follicular dilatations: First laboratory experiments using colonies of *Lutzomyia flaviscutellata* (Mangabeira) and *L. furcata* (Mangabeira) (Diptera: Psychodidae), *Bull. Entomol. Res.* **74**:641–646.
136. Rhodendorf, B., 1964, The historical development of Diptera, *Trudy. Paleontol. Inst.* **100**:1–300 (in Russian) [English translation: *The Historical*

Development of Diptera University of Alberta Press, Alberta, 1974. 360 pp.].

137. Rioux, J.A., Lanotte, G., Croset, H., Houin, R., Guy, Y., and Dedet, J.P., 1972, Écologie des leishmanioses dans le sud de la France. 3. Réceptivité comparée de *Phlebotomus ariasi* Tonnoir, 1921 et *Rhipicephalus turanicus* Pomerancev et Matikasvili, 1940 vis-à-vis de *Leishmania donovani* (Laveran et Mesnil, 1903), *Ann. Parasitol. Hum. Comp.* **47**:147–157.
138. Rioux, J.A., Killick-Kendrick. R., Leaney, A.J., Young, C.J., Turner, D.P., Lanotte, G., and Bailly, M., 1979, Écologie des leishmanioses dans le sud de la France. 11. La leishmaniose viscérale canine: Succès de la transmission expérimentale "chien → phlebotome → chien" par la piqûre de *Phlebotomus ariasi* Tonnoir, 1921, *Ann. Parasitol. Hum. Comp.* **54**:27–407.
139. Rodriguez, M.J.D., 1969, Leishmaniasis muco-cutanea en la Province de Pichincha, *Rev. Ecuat. Hyg. Med. Trop.* **26**:3–7.
140. Rutledge, L.C., and Ellenwood, D.A., 1975a, Production of phlebotomine sandflies on the open forest floor in Panama: The species complement, *Environ. Entomol.* **4**:71–77.
141. Rutledge, L.C., and Ellenwood, D.A., 1975b, Production of phlebotomine sandflies on the open forest floor in Panama: Hydrologic and physiologic relations, *Environ. Entomol.* **4**:78–82.
142. Rutledge, L.C., and Ellenwood, D.A., 1975c, Production of phlebotomine sandflies on the open forest floor in Panama: Phytologic and edaphic relations, *Environ. Entomol.* **4**:83–89.
143. Rutledge, L.C., and Mosser, H.I., 1972, Biology of immature sandflies (Diptera: Psychodidae) at the bases of trees in Panama, *Environ. Entomol.* **1972**:300–309.
144. Ryan, L., Silveira, F.T., Lainson, R., and Shaw, J.J., 1984, Leishmanial infections in *Lutzomyia longipalpis* and *Lu. antunesi* (Diptera: Psychodidae) on the island of Marajó, Pará State, Brazil, *Trans. R. Soc. Trop. Med. Hyg.* **78**:547–548.

144a. Ryan, L., Phillips, A., Milligan, P., Lainson, R., Molyneux, D.H., and Shaw, J.J., 1986, Separation of female *Psychodopygus wellcomei* and *P. Complexa* (Diptera: Psychodidae) by cuticular hydrocarbon analysis, *Acta Tropica.* **43**:85–89.

145. Sacks, D.L., and Perkins, P.V., 1984, Identification of an infective stage of *Leishmania* promastigotes, *Science* **223**:1417–1449.
146. Sanderson, M.W., and Farr, T.H., 1960, Amber with insect and plant inclusions from the Dominican Republic, *Science* **131**:1313.
147. Sapunar, J., Díaz, M., Wolf, R., and Tello, P., 1980, Leishmaniasis cutánea, *Bol. Chil. Parasitol.* **35**:25–28.

147a. Saravia, N.G., Holguín, A.F., McMahon-Pratt, D., and D'Allesandro, A., 1985, Mucocutaneous leishmaniasis in Colombia: *Leishmania braziliensis* subspecies diversity, *Am. J. Trop. Med. Hyg.* **34**:714–720.

148. Schlein, Y., and Warburg, A., 1985, Feeding behavior, midgut distension and ovarian development in *Phlebotomus papatasi* (Diptera: Psychodidae), *J. Insect Physiol.* **31**:45–51.

148a. Schlein, Y., and Warburg, A., 1986, Phytophagy and the feeding cycle of *Phlebotomus papatasi* (Diptera: Psychodidae) under experimental conditions, *J. Med. Entomol.* **23**:11–15.

149. Schlein, Y., Yuval, B., and Warburg, A., 1984, Aggregation pheromone released from the palps of feeding female *Phlebotomus papatasi* (Psychodidae), *J. Insect Physiol.* **30:**153–156.
150. Schnur, L.F., Chance, M.L., Ebert, F., Thomas, S.C., and Peters, W., 1981, The biochemical and serological taxonomy of visceralizing *Leishmania, Ann. Trop. Med. Parasitol.* **75:**251–253.
151. Schnur, L.F., Walton, B.C., and Bogaert-Díaz, H., 1983, On the identity of the parasite causing diffuse cutaneous leishmaniasis in the Dominican Republic, *Trans. R. Soc. Trop. Med. Hyg.* **77:**756–762.
151a. Scorza, J.V., and Añez, N., 1984, Transmissión experimental de *Leishmania garnhami* al hamster por la picadura de *Lutzomyia townsendi, Rev. Cubana Med. Trop.* **36:**139–145.
152. Scorza, J.V., and Delgado, O., 1982, Morfometria amastigota y desarrollo de cuatro aislados de *Leishmania mexicana pifanoi* de Venezuela en *Lutzomyia townsendi, Mem. Inst. Oswaldo Cruz* **77:**217–227.
153. Scorza, J.V., Valera, M., de Scorza, C., Carnevali, M., Moreno, E., and Lugo-Hernandez, A., 1979, A new species of *Leishmania* parasite from the Venezuelan Andes region, *Trans. R. Soc. Trop. Med. Hyg.* **73:**293–298.
154. Shaw, J.J., 1981, The behaviour of *Endotrypanum schaudinni* (Kinetoplastida: Trypanosomatidae) in species of laboratory-bred neotropical sandflies (Diptera: Psychodidae) and its influence on the classification of the genus *Leishmania,* in: E.U. Canning (ed.), *Parasitological Topics, A Presentation Volume to P.C. Garnham on the Occasion of His 80th Birthday,* Allen Press, Lawrence, Kansas, pp. 48–53.
155. Shaw, J.J., and Lainson, R., 1968, Leishmaniasis in Brazil. II. Observations on enzootic rodent leishmaniasis in the lower Amazon region—The feeding habits of the vector, *Lutzomyia flaviscutellata* in reference to man, rodents and other animals, *Trans. R. Soc. Trop. Med. Hyg.* **62:**396–405.
156. Sherlock, I.A., 1964, Notas sôbre a transmissão da leishmaniose visceral no Brasil, *Rev. Bras. Malariol. Doen. Trop.* **16:**19–26.
157. Spithill, T.W., and Grumont, R.J., 1984, Identification of species, strains and clones of *Leishmania* by characterization of kinetoplast DNA minicircles, *Mol. Biochem. Parasitol.* **12:**217–236.
158. Strangways-Dixon, J., and Lainson, R., 1966, The epidemiology of dermal leishmaniasis in British Honduras. Part III. The transmission of *L. mexicana* to man by *Phlebotomus pessoanus* with observations on the development of the parasite in different species of *Phlebotomus, Trans. R. Soc. Trop. Med. Hyg.* **60:**192–201.
159. Telford, S.R., Herrer, A., and Christensen, H.A., 1972, Enzootic cutaneous leishmaniasis in eastern Panama. III. Ecological factors relating to the mammalian hosts, *Ann. Trop. Med. Parasitol.* **66:**173–179.
160. Tesh, R.B., Chaniotis, B.N., Aronson, M., and Johnson, K.M., 1971, Natural host preferences of Panamanian phlebotomine sandflies as determined by precipitin test, *Am. J. Trop. Med. Hyg.* **20:**150–156.
161. Tesh, R.B., Chaniotis, B.N., Carrera, B.R., and Johnson, K.M., 1972, Further studies on the natural host preferences of Panamanian phlebotomine sandflies, *Am. J. Epidemiol.* **95:**88–93.
162. Theodor, O., 1948, Classification of the Old World species of the subfamily Phlebotominae (Diptera: Psychodidae), *Bull. Entomol. Res.* **39:**85–115.

163. Theodor, O., 1965, On the classification of American Phlebotominae, *J. Med. Entomol.* **2**:171–97.
164. Tibayrenc, M., Carriou, M.L., Corneau, B., and Pajot, F.-X., 1980, Étude allozymique chez *Lutzomyia umbratilis* (Diptera: Psychodidae), vecteur de la leishmaniose en Guyane Française, *Cah. ORSTOM Ser. Entomol. Med. Parasitol.* **18**:67–70.
165. Tikasingh, E.A., 1975, Observations on *Lutzomyia flaviscutellata* (Mangabeira) (Diptera: Psychodidae), a vector of enzootic leishmaniasis in Trinidad, West Indies, *J. Med. Entomol.* **12**:228–232.
166. Van Handel, E., 1972, The detection of nectar in mosquitoes, *Mosq. News* **32**:589–591.
167. Van Handel, E., 1984, Metabolism of nutrients in the adult mosquito, *Mosq. News* **44**:573–579.
168. Walker, T.J., 1980, Mixed oviposition in individual females of *Gryllus firmus:* Graded proportions of fast-developing and diapause eggs, *Oecologia* **47**:291–298.
168a. Walters, L.L., Modi, G.B., Tesh, R.B., and Burrage, T., 1987, Host–parasite relationship of *Leishmania mexicana mexicana* and *Lutzomyia abonnenci* (Diptera: Psychodidae), *Am. J. Trop. Med. Hyg.* **36** (in press).
169. Ward, R.D., 1976, The immature stages of some phlebotomine sandflies from Brazil (Diptera: Psychodidae), *Syst. Entomol.* **1**:227–240.
170. Ward, R.D., 1977a, The colonization of *Lutzomyia flaviscutellata* (Diptera: Psychodidae), a vector of *Leishmania mexicana amazonensis* in Brazil, *J. Med. Entomol.* **14**:469–476.
171. Ward, R.D., 1977b, New World leishmaniasis: A review of the epidemiological changes in the last three decades, in: *Proceedings of the XV International Congress of Entomology, Washington,* pp. 505–522.
172. Ward, R.D., and Ready, P.D., 1975, Chorionic sculpturing in some sandfly eggs (Diptera: Psychodidae), *J. Entomol.* **50**:127–134.
173. Ward, R.D., Lainson, R., and Shaw, J.J., 1973a, Further evidence of the role of *Lutzomyia flaviscutellata* (Mangabeira) as the vector of *Leishmania mexicana amazonensis* in Brazil, *Trans. R. Soc. Trop. Med. Hyg.* **67**:608–609.
174. Ward, R.D., Shaw, J.J., Lainson, R., and Fraiha, H., 1973b, Leishmaniasis in Brazil. VIII. Observations on the phlebotomine fauna of an area highly endemic for cutaneous leishmaniasis in the Serra do Carajás, Pará State, *Trans. R. Soc. Trop. Med. Hyg.* **67**:174–183.
175. Ward, R.D., Lainson, R., and Shaw, J.J., 1977, Experimental transmissions of *Leishmania mexicana amazonensis* Lainson and Shaw, between hamsters by the bite of *Lutzomyia flaviscutellata* (Mangabeira), *Trans. R. Soc. Trop. Med. Hyg.* **71**:265–266.
176. Ward, R.D., Ribeiro, A.L., Ready, P.D., and Murtagh, A., 1983, Reproductive isolation between different forms of *Lutzomyia longipalpis* (Lutz and Neiva) (Diptera: Psychodidae), the vector of *Leishmania donovani chagasi* Cunha and Chagas and its significance to kala-azar distribution in South America, *Mem. Inst. Oswaldo Cruz* **78**:269–280.
177. White, G.B., and Killick-Kendrick, R., 1976, Polytene chromosomes of the sandfly *Lutzomyia longipalpis* and the cytogenetics of Psychodidae in relation to other Diptera, *J. Entomol. A* **50**:187–196.

178. WHO, 1984, The Leishmaniases, Report of a WHO Expert Committee, Technical Report Series No. 701, World Health Organization. 140 pp.
179. Wijers, D.J.B., and Linger, R., 1967, Man-biting sandflies in Surinam (Dutch Guiana): *Phlebotomus anduzei* as a possible vector of *Leishmania braziliensis, Ann. Trop. Med. Parasitol.* **60:**501–508.
180. Wilkes, T.J., Ready, P.D., Lainson, R., and Killick-Kendrick, R., 1984, Biting periodicities of nulliparous and parous females of *Psychodopygus wellcomei, Trans. R. Soc. Trop. Med. Hyg.* **78:**846–847.
181. Williams, P., 1966, Experimental transmission of *Leishmania mexicana* by *Lutzomyia cruciata, Ann. Trop. Med. Parasitol.* **60:**365–372.
182. Williams, P., 1970, Phlebotomine sandflies and leishmaniasis in British Honduras (Belize), *Trans. R. Soc. Trop. Med. Hyg.* **64:**317–368.
183. Williams, P., 1976, The phlebotomine sandflies (Diptera, Psychodidae) of caves in Belize, Central America, *Bull. Entomol. Res.* **65:**601–614.
184. Williams, P., 1983, The identity of the sandfly that first experimentally transmitted a neotropical *Leishmania, Trans. R. Soc. Trop. Med. Hyg.* **77:**489–491.
185. Williams, P., and Coelho, M.V., 1978, Taxonomy and transmission of *Leishmania,* in: W.H.R. Lumsden, R. Muller, and J.R. Baker (eds.), *Advances in Parasitology,* Vol. 16, Academic Press, New York, pp. 1–42.
186. Wirth, D.F., and McMahon-Pratt, D., 1982, Rapid identification of *Leishmania* species by specific hybridization of kinetoplast DNA in cutaneous lesions, *Proc. Natl. Acad. Sci. USA* **79:**6999–7003.
187. Woods, C.A., 1981, Last endemic mammals in Hispaniola, *Oryx* **16:**146–152.
188. Yarbuh, A.L., and Scorza, J.V., 1982, Ensayos metodólogicos para la investigación de reservoirs de *Leishmania* spp. en los Andes Venezolanos, *Mem. Inst. Oswaldo Cruz* **77:**367–384.
189. Young, D.G., 1972, Phlebotomine sand flies from Texas and Florida (Diptera: Psychodidae), *Florida Entomol.* **55:**61–64.
190. Young, D.G., 1979, A review of the bloodsucking psychodid flies of Colombia (Diptera: Phlebotominae and Sycoracinae), University of Florida Agricultural Experimental Station Technical Bulletin 806. 266pp.
191. Young, D.G., and Arias, J.R., 1984, The *microps* group of *Lutzomyia* França with descriptions of two new species from South America (Diptera: Psychodidae: Phlebotominae), *Mem. Inst. Oswaldo Cruz* **79:**425–431.
192. Young, D.G., and Perkins, P.V., 1984, Phlebotomine sand flies of North America (Diptera: Psychodidae), *Mosq. News* **44:**263–304.
193. Young, D.G., Pérez, J.C., and Romero, G., 1985, New records of phlebotomine sand flies from Peru with a description of *Lutzomyia oligodonta* n. sp. from the Rimac Valley (Diptera: Psychodidae), *Int. J. Entomol.* **27:**136–146.
193a. Young, D.G., Morales, A., Kreutzer, R.D., Alexander, J.B., Corredor, A., and Tesh, R.B., 1987, Isolations of *Leishmania braziliensis* from cryopreserved Colombian sand flies (Diptera: Psychodidae), *J. Med. Entomol.* (submitted).
194. Zeledón, R., and Alfaro, M., 1973, Isolation of *Leishmania braziliensis* from a Costa Rican sandfly and its possible use as a human vaccine, *Trans. R. Soc. Trop. Med. Hyg.* **67:**416–417.

195. Zeledón, R., McPherson, B., and Ponce, C., 1977, Isolation of *Leishmania braziliensis* from a wild rodent in Costa Rica, *Am. J. Trop. Med. Hyg.* **26**:1044–1045.
195a. Zeledón, R., Murillo, J., and Gutiérrez, H., 1985, Flebótomos antropófilos y leishmaniasis cutánea en Costa Rica, *Bol. Ofic. Sanit. Panamericana.* **99**:163–172.
196. Zeledón, R., Ponce, C., and Murillo, J., 1979, *Leishmania herreri* sp. n. from sloths and sandflies of Costa Rica, *J. Parasitol.* **65**:275–279.
197. Zeledón, R., Macaya, G., Ponce, C., Chaves, F., Murillo, J., and Bonilla, J.A., 1982, Cutaneous leishmaniasis in Honduras, Central America, *Trans. R. Soc. Trop. Med. Hyg.* **76**:276.
198. Zeledón, R., Murillo, J., and Gutierrez, H., 1984, Observaciones sobre la ecologia de *Lutzomyia longipalpis* (Lutz and Neiva, 1912) y posibilidades de existencia de leishmaniasis visceral en Costa Rica, *Mem. Inst. Oswaldo Cruz* **79**:455–459.
199. Zimmerman, J.H., Newson, H.D., Hooper, G.R., and Christensen, H.A., 1977, A comparison of the egg surface structure of six anthropophilic phlebotomine sandflies *(Lutzomyia)* with the scanning electron microscope (Diptera: Psychodidae), *J. Med. Entomol.* **13**:574–579.

3
Whitefly Transmission of Plant Viruses

James E. Duffus

Introduction

Whitefly-transmitted disease agents cause significant losses throughout the world. Although not considered as important as aphids on a worldwide basis, they are responsible for the natural spread of a large number of economically important diseases in the tropical and subtropical areas. Recent years have shown an increase in losses in wide areas north and south of the tropics, approaching areas of intensive agricultural production such as the southern United States, Jordan, and Israel (26, 28, 32, 43). These areas are in the apparently increasing range of *Bemisia tabaci* (Gennadius), the most intensively studied whitefly vector. Recent years have shown, if not an absolute increase, at least an increase in the awareness of disease losses caused by two other whitefly species, *Trialeurodes vaporarium* (Westwood) and *T. abutilonea* (Hald.), in temperate areas of the United States, Europe, Australasia, and Asia (41, 42, 58, 60, 86, 102).

The importance of the whitefly-transmitted agents becomes apparent when the crops affected by the diseases induced by them are reviewed. Serious losses are induced on cassava, cotton, cowpea, bean, tobacco, soybean, tomato, squash, melon, watermelon, lettuce, sugar beet, carrot, peppers, sweet potato, cucumber, and papaya (1, 3, 4, 17, 19, 25, 29, 30, 32, 38, 43, 46). Some 70 or more diseases have been reported to be induced by the feeding of infectious whiteflies. The relationships between these diseases, often poorly described, are not well established and in many instances they probably represent diseases induced by the same agent or strains of that agent.

Insects have a surprisingly long history and date back about 250 million years to the Paleozoic era. Yet it was not until 1930 that Kirkpatrick (57)

James E. Duffus, U.S. Department of Agriculture, Agricultural Research Service, Salinas, California 93905, USA.

first found that cotton leaf curl virus was transmitted by the whitefly *Bemisia tabaci*. This came some 34 years after the report by Takami (83) that a leafhopper was responsible for the transmission of the agent responsible for rice dwarf and 16 years after Allard (2) experimented with aphid transmission of tobacco mosaic virus and concluded that *Myzus persicae* was a vector—but the insect vector relationships with tobacco mosaic are still obscure, which illustrates the complex nature of the transmission of disease agents by insects.

The earliest investigators of plant viruses sucked juice of diseased plants into glass capillaries or syringes and injected these into the midribs of healthy plants. In the 1920s, inoculations were commonly made by pricking through drops of juice into healthy plants. These techniques obviously affected the thinking of early workers in regard to the mechanical transmission mode of plant disease agent transmission by insects—the disease agents were carried as superficial contaminants by the stylets and introduced mechanically into plants as though by a pin. However, such factors as vector specificity, ranging from degree of efficiency to the inability of some species to transmit some disease agents, the apparent lack of transmission of some disease agents, and the ability of vectors to retain some agents for much longer periods than others stimulated an early experimental examination of the transmission process.

The earliest work with whitefly transmission of cotton leaf curl, tobacco leaf curl, and cassava mosaic were all of the persistent type, which until recent years has given the impression that whitefly-transmitted disease agents were somehow different, with greater vector dependence than the viruses transmitted by other vectors (82).

A number of recent reviews have been published on whitefly-transmitted disease agents (10, 31, 33, 63, 90). The whitefly-transmitted diseases in general have been characterized on the basis of the transmission by whiteflies and activity of the agents on host plants, such as symptoms, cross protection, and host range. Costa (33) grouped the whitefly-transmitted diseases into three main categories based on symptoms: (1) mosaic diseases, (2) leaf curl diseases, and (3) the yellowing diseases. Muniyappa (63) grouped the diseases along similar lines: (1) yellow mosaic diseases, (2) yellow vein mosaics, (3) leaf curl diseases, and (4) mosaic and other types of virus diseases. Bird and Maramorosch (10) used the term "rugaceous diseases" to designate diseases of uncertain etiology transmitted by *Bemisia tabaci* and characterized by symptoms on their hosts such as malformations, leaf curl, enations, and yellow mosaic. Only recently has research given enough evidence to distinguish the whitefly-transmitted viruses on the basis of particle morphology, serology, and biochemical methods.

A compilation of data on whitefly-transmitted viruses at this point would suggest at least seven groups of viruses differing in type of virus particle, symptom type, and vector relationships. Since the data on the some 70 diseases reported to be transmitted by whiteflies is so limited, the number of groups is undoubtedly much larger.

Only one of the established groups of plant viruses, the geminiviruses, has been reported and described as being transmitted by whiteflies. The group also has important members transmitted by leafhoppers. Since the natural method of transmission is one of the eight criteria used in virus classification by the Subcommittee on Virus Nomenclature, it is difficult to place newly described whitefly-transmitted viruses in the "older" virus groups until the vector relationships, purification, serology, and biochemistry have been done and compared with members of the group.

In the tentative categorization of the whitefly-transmitted viruses discussed here, it should be remembered that much information in regard to particle type, serology, and insect transmission is not yet available. Grouping of the viruses was based on similarities of the viruses on the characters that were available.

It is apparent that whitefly-transmitted viruses are much more diverse than originally thought. Virtually all types of transmission occur with the viruses and whitefly vectors, ranging from nonpersistent, semipersistent, and persistent, to a degree of active interaction between the virus and vector in some of the geminiviruses. Symptoms are also as varied as in the aphid-transmitted viruses and so perhaps are the virus types that are transmitted by whiteflies.

Based on the data used for classifying viruses, only three criteria in the whitefly literature are useful in separating the viruses: the structure of the virus particles, activities in the plant, such as symptom type and cytology, and methods of transmission.

Definitive members of the geminivirus group by particle morphology or serology all show symptoms of the variegation or leaf curl type on susceptible plants and are persistent in their vectors.

Members of the "closterovirus-like" group by particle morphology all show symptoms on susceptible plants of the yellow vein or interveinal yellowing type. Three of the five possible members of this group have semipersistent relationship with their vectors. The persistence of cucumber yellows, or *Diodia* yellow vein viruses, has not yet been determined. The viruses are not mechanically transmitted.

Members of the "carlavirus-like" group, with particles 650–750 nm, show mosaic or ringspot symptoms, are nonpersistent in whitefly vectors, and are mechanically transmitted.

Only one member is described with particle morphology similar to the potyvirus group (flexuous rods 800–950 nm). Sweet potato mild mottle virus has been mechanically transmitted, but has undetermined insect relationships (53).

One recently described virus, tomato necrotic dwarf, has particle morphology and chemistry similar to the nepovirus or nepolike viruses (59). The virus has three distinct isometric components approximately 30 nm in diameter. Infected plants show mosaic or ringspot symptoms and the virus is transmitted mechanically and in a nonpersistent manner by whitefly vectors.

A unique virus, cucumber yellow vein virus with rod-shaped particles containing double-stranded DNA, forms another distinctive group (26, 69). If this work is confirmed and the virus indeed contains a DNA genome, this is the fourth rod-shaped whitefly-transmitted group. This virus has particles 740–900 nm long, is semipersistent in its whitefly vectors, and is mechanically transmitted.

Another possible group of whitefly-transmitted viruses is made up of those with interveinal yellowing and yellow vein symptoms transmitted in a persistent manner. No particles have been visualized in this group, but none of the viruses have been mechanically transmitted. They are distinct from the rod-shaped viruses in their persistence and from the geminiviruses in regard to symptoms. They have symptoms and insect vector relationships not unlike the luteoviruses.

Geminiviruses

Geminiviruses have small, 18-nm icosahedral nucleoprotein particles, which occur in pairs. Most possess a single major coat protein subunit with a molecular weight in the range $(2.7–3.4) \times 10^4$. The nucleic acid is a circular molecule of single-stranded DNA with molecular weight of around 8.0×10^5. The viruses are transmitted by leafhoppers or whiteflies in a persistent manner. In the persistent transmission relationship, virus is acquired by insects feeding on an infected plant, and usually from its phloem, and passes through the gut wall to the hemolymph and salivary glands. It must pass through the salivary glands and into the insect's saliva. There is a positive correlation between the duration of both the acquisition feeding and the inoculation feeding periods and the probability of transmissions. A latent period, although it may be brief and poorly defined in terms of a minimum, occurs between the acquisition feed and the vector becoming infective. The period of retention of the persistent viruses is relatively long, a number of weeks to life, and the insect retains this ability through a molt.

Geminivirus infections are associated with particulate nuclear inclusions including electron-dense condensed fibrillar rings, and the appearance of virus particles in aggregates. Squash leaf curl virus (52) shows infection of leaf phloem parenchyma, xylem parenchyma, vein border parenchyma, and mesophyll cells near veins. Bean golden mosaic and *Euphorbia* mosaic showed virus particles and cytopathic effects only in phloem parenchyma and young sieve elements, which is somewhat unexpected, since these two viruses are also mechanically transmitted (55, 56).

It is rare among the insect-transmitted viruses to be persistent in the vector and to be mechanically transmitted. Persistence in the vector usually is associated with viruses with high tissue specificity involving the phloem and these viruses generally do not involve epidermal or mesophyll tissues. The bright yellow variegated symptoms associated with the mechanically transmitted geminiviruses are unique in appearance and the distribution of virus. The apparent landing behavior of whiteflies, attraction to yellow color, short-sightedness, and passive landing behavior seem to be especially adaptive to vectoring by whiteflies (22).

Members of the geminivirus group and possible members, where tested, have fairly similar transmission characteristics, except perhaps for persistence (Tables 3.1 and 3.2). The viruses are acquired by whiteflies in as short a period as 10 min, but efficiency of transmission increases with feeding periods up to 24 hr. Inoculation periods as short as 10 min are adequate for infection, but transmission efficiency increases with longer inoculation periods. Latent periods between acquisition of virus and the ability of the insects to infect plants varied from 4 to 21 hr, but these differences may reflect transmission efficiency differences rather than real differences between the viruses. Where adequately tested, *Euphorbia* mosaic, bean golden mosaic, tobacco leaf curl, squash leaf curl, and other whitefly-transmitted geminiviruses were retained by the vectors for 20 or more days. Several of the viruses in the group have been shown to be acquired in the nymphal stages and to transmit the virus to healthy plants as they emerge as adult whiteflies from their pupal cases.

A unique phenomenon in plant virus–insect relationships is that termed "periodic acquisition" by Cohen and Harpaz (24). Whiteflies with tomato yellow leaf curl virus (TYLCV) progressively lose infectivity and most insects do not transmit virus subsequent to 10 days after completion of acquisition feeding. Although progressive loss of inoculativity of the vector cannot be prevented by repeated or prolonged acquisition on an infected plant, inoculativity could be restored only when it first completely ceased to transmit the virus. An antiviral factor termed "periodic acquisition related factor" (PARF) was found in homogenates of whiteflies carrying TYLCV. Feeding of whiteflies through parafilm membranes on PARF prior to or after virus acquisition resulted in reduced ability of the insects to acquire and transmit the virus. A new factor, PARFa, was found to be of proteinaceous nature and it suppressed TYLCV inoculativity of the whiteflies only when injected into them (20, 21, 25).

A similar phenomenon was observed with tomato yellow mosaic virus (92) and may occur with other whitefly-transmitted geminiviruses. The periodic acquisition phenomenon indicates more than a passive role for certain geminiviruses in their whitefly vectors.

TABLE 3.1. Whitefly-transmitted viruses of the geminivirus group.

Virus	Particles observed	Serological relationship	Vector	Type of persistence[a]	Transmission through molt	Symptoms[b]	Mechanical transmission	Reference
Abutilon mosaic	Gemini	—	Bemisia	P	—	V	Yes	35, 36, 70
African cassava mosaic	Gemini	Gemini	Bemisia	P	—	V	Yes	17, 72, 81
Bean golden mosaic	Gemini	Gemini	Bemisia	P	—	V	Yes	12, 30, 49, 50
Cotton leaf crumple	Gemini	—	Bemisia	P	No	C	No	3, 15, 40, 87
Croton yellow vein mosaic	Gemini	Gemini	Bemisia	—	—	YV	—	65, 90
Euphorbia mosaic	Gemini	Gemini	Bemisia	P	Yes	V	Yes	13, 34
Horsegram yellow mosaic	Gemini	—	Bemisia	P	—	V	—	63
Melon leaf curl	Gemini	Gemini	Bemisia	P	—	C	Yes	46
Tobacco leaf curl	Gemini	Gemini	Bemisia	P	—	C, V	Yes	80, 101
Tomato golden mosaic	Gemini	—	Bemisia	—	—	V	Yes	79
Tomato yellow leaf curl	Gemini	Gemini	Bemisia	P	Yes	C	Yes	24, 27
Tomato yellow mosaic	Gemini	—	Bemisia	P	—	C, V	Yes	37, 92
Squash leaf curl	Gemini	Gemini	Bemisia	P	—	C	Yes	29

[a]Type of persistence: P, persistent.
[b]Type of symptoms: C, leaf curl; V, varigation; YV, yellow vein.

TABLE 3.2. Whitefly-transmitted viruses with possible relationships to the geminivirus group.

Virus	Particles observed	Serological relationship	Vector	Type of persistence[a]	Transmission through molt	Symptoms[b]	Mechanical transmission	Reference
Acalypha yellow mosaic	—	—	Bemisia	—	—	C, V	—	18
Ageratum yellow vein mosaic	—	—	Bemisia	P	—	C, YV	No	48, 65, 90
Balsam leaf curl	—	—	Bemisia	—	—	C	—	96
Cape-gooseberry leaf curl	—	—	Bemisia	—	—	C	—	67
Cowpea yellow mosaic	—	—	Bemisia	—	—	V	—	1, 68
Cotton leaf curl	—	—	Bemisia	P	Yes	C	—	57, 84
Dolichos lab-lab yellow mosaic	—	—	Bemisia	—	—	V	—	71
Hollyhock yellow mosaic	—	—	Bemisia	—	—	V	—	76, 77
Jacquemontia yellow mosaic	—	—	Bemisia	—	—	C, V	—	13
Jatropha leaf curl	—	—	Bemisia	—	—	C	—	66
Jatropha yellow mosaic	—	—	Bemisia	—	Yes	C, V	No	8, 13
Jute yellow mosaic	—	—	Bemisia	—	—	V	—	14, 93
Hibiscus leaf curl	—	—	Bemisia	—	—	C	—	90, 98
Lupin leaf curl	—	—	Bemisia	—	—	C	—	94
Merremia yellow mosaic	—	—	Bemisia	—	—	V	—	12, 13
Potato leaf curl	—	—	Bemisia	—	—	C	—	101
Rhynchosia yellow mosaic	—	—	Bemisia	—	Yes	V	—	9, 13
Sesamum leaf curl	—	—	Bemisia	—	—	C	—	73
Soapwort leaf curl	—	—	Bemisia	—	—	C	—	95
Wissadula mosaic	—	—	Bemisia	P	Yes	V	—	74
Zinnia leaf curl	—	—	Bemisia	—	—	C	—	62, 97

[a]Type of persistence: P, persistent.
[b]Type of symptoms: C, leaf curl; V, varigation; YV, yellow vein.

Carlaviruses

Carlaviruses have virions that are slightly flexuous rods, potyviruses are generally more flexuous, and closteroviruses much more flexuous, with length of about 620–690 nm.

The molecular weight range of the protein subunits is $(3.1\text{–}3.6) \times 10^4$. The nucleic acid is RNA with estimated molecular weight of $(2.3\text{–}2.6) \times 10^6$.

The viruses generally cause mild diseases, latent infections, mild mottle, or vein clearing symptoms. Carlaviruses are mechanically transmitted and many have aphid vectors. They are transmitted by aphids in the nonpersistent manner.

In the nonpersistent relationship, the acquisition and inoculation threshold periods are but a few seconds in duration and the probability of acquisition increases if the insects have fasted for a short time before making an acquisition probe into diseased tissue. As the duration of the acquisition probe or feeding penetration increases, there is a decrease in the probability of acquisition of virus.

Once an insect has acquired a charge of virus from the brief acquisition probe, the subsequent activities of the vector are critical if transmission is to occur. If the insect has the opportunity to make a rapid series of brief inoculation probes, successive transmissions of the virus, although somewhat limited in number, are possible. The probability, however, of transmission tends to decrease with each successive probe.

The period of retention of the nonpersistent viruses is a short time, a number of minutes to hours.

The two whitefly-transmitted viruses with properties like the carlaviruses that have been studied in regard to insect transmission properties are similar to each other but differ slightly from nonpersistently transmitted aphid-transmitted viruses (Table 3.3). Both cowpea mild mottle virus (CMMV) (64) and tomato pale chlorosis disease virus (TPCDV) (23) can be acquired in minutes (10–15 min) and inoculated into healthy plants in 5–15 min. Higher inoculation rates were obtained by acquisition feedings of from ½ to 1 hr, but not with longer acquisition feedings.

Preacquisition fasting had no influence on transmission. CMMV was retained by the vector for 1 hr and TPCDV for about 6 hr.

Closteroviruses

Closteroviruses have particles that are very flexuous filamentous rods with lengths of 600 to about 2000 nm. The viruses cause yellowing, vein yellowing, and phloem necrosis and have moderate host ranges. Some of the members are mechanically transmitted with difficulty. Most of the members are aphid-transmitted in a semipersistent manner. In the semipersis-

TABLE 3.3. Whitefly-transmitted viruses with possible relationships to the carlaviruses.

Virus	Particles observed	Serological relationship	Vector	Type of persistence[a]	Transmission through molt	Symptoms[b]	Mechanical transmission	Reference
Cowpea mild mottle	Rod 650–700 nm	Carlavirus	Bemisia	NP	—	M	Yes	16, 54, 64
Tomato pale chlorosis	Rod 700 nm	—	Bemisia	NP	—	M, VC	Yes	23
Jasmine chlorotic ringspot	Rod 700–750 nm	—	Bemisia	—	—	M, RS	—	5, 99

[a]Type of persistence: NP, nonpersistent.
[b]Type of symptoms: M, mottle; RS, ringspot; VC, vein clearing.

TABLE 3.4. Whitefly-transmitted viruses with possible relationships to the closteroviruses.

Virus	Particles observed	Serological relationship	Vector	Type of persistence[a]	Transmission through molt	Symptoms[b]	Mechanical transmission	Reference
Abutilon yellows	—	—	Trialeurodes abutilonea	SP	—	Y, YV	No	J.E. Duffus (unpublished)
Beet pseudo yellows	Rod 1500 nm	—	Trialeurodes vaporariorum	SP	—	Y	No	41, 42, 44
Cucumber yellows	Rod 1000 nm	—	Trialeurodes vaporariorum	—	—	Y	—	102
Diodia yellow vein	Rod 1500 nm	—	Trialeurodes abutilonea	—	—	YV	—	59
Lettuce infectious yellows	Rod 2000 nm	—	Bemisia	SP	—	Y	No	43, 45, 46

[a]Type of persistence: SP, semipersistent.
[b]Type of symptoms: Y, interveinal yellowing; YV, yellow vein.

tent relationship there is a positive correlation between the length of the acquisition feeding period and the probability of virus acquisition. There is also a positive correlation between duration of the inoculation feeding period and successful transmission, although the inoculation efficiency curve generally reaches a maximum within a shorter time interval than that of acquisition.

There does not appear to be any accumulation of inoculative capacity after an acquisition feeding period has ended. The greatest probability of transmission occurs upon leaving the virus source. The likelihood of transmission then gradually decreases to zero with time. The rate at which inoculativity decreases in either feeding or fasting insects, while influenced by the ambient temperature, is measurable in periods of hours rather than in minutes as in the nonpersistent relationship, or days as in the case of the persistent type of transmission. The acquisition of virus depends upon feeding rather than probing.

The closteroviruses, in general, are associated with intracellular inclusions including characteristic vesicles. The coat proteins of those viruses examined for molecular weights range between 2.3×10^3 and 2.7×10^4 and the nucleic acid is RNA estimated at $(2–4.5) \times 10^6$.

Five whitefly-transmitted viruses have some characteristics similar to the aphid-transmitted closteroviruses, including *Abutilon* yellows, beet pseudo yellows, cucumber yellows, *Diodia* yellow vein, and lettuce infectious yellows virus (41–43, 45, 46, 58, 60, 86, 102) (Table 3.4). The characteristics of this group of viruses include symptoms of the yellowing type, interveinal yellowing, and brittleness of affected leaves, and/or vein yellowing symptoms. Virus particles are long, flexuous rods 1000–2000 nm long and none have been transmitted mechanically. The viruses of the group are transmitted by three distinct species of whitefly, *Trialeuroides vaporariorum, T. abutilonea,* and *Bemisia tabaci,* all in a semipersistent manner. Each virus seems to be transmitted by only one whitefly species. Cytopathological examinations of diseased plants of those diseases studied revealed the presence of flexuous, rod-shaped particles in the cytoplasm of phloem parenchyma cells and sieve elements. The particles were associated with membranous vesicles containing fibrils.

Those viruses studied can be acquired by nonviruliferous whiteflies in a period as short as 10 min. There is a positive correlation between the length of the acquisition feeding period and the duration of the inoculation feeding period and the probability of virus transmission. The viruses are retained by the whiteflies for periods of days, ranging from 3 days for LIYV to 7 days for PSYV.

Viruses with Possible Relationships to the Potyviruses

The potyviruses have filamentous particles measuring 680–900 nm $\times$ 11 nm wide. They contain single-stranded RNA of $(3.0–3.5) \times 10^6$ molecular weight and a single protein of $(3.2–3.4) \times 10^4$). The viruses induce the formation of cytoplasmic inclusions, such as pinwheels and laminated aggregates.

The viruses are transmitted in the nonpersistent manner by aphids and by mechanical transmission.

Sweet potato mild mottle virus (SPMMV) resembles potyviruses in having filamentous particles 800–900 nm long with one protein of 3.7×10^4 molecular weight and single-stranded RNA (53) (Table 3.5). It induces the formation of cytoplasmic pinwheel inclusions, but is unrelated to 14 viruses of the potyvirus group.

The vector relationships of SPMMV have not been determined, but in preliminary tests it was necessary to expose test plants for at least 5 days to ensure transmission. The author suggested that the virus was not stylet-borne. If the virus is shown to be transmitted in a manner other than stylet-borne, it should probably be excluded from a close relationship to the group.

Viruses with Possible Relationships to the Luteoviruses

Luteoviruses have small, isometric virions with icosahedral symmetry and a diameter of about 25 nm. Sedimentation coefficients range from 115 to 127S. Most viruses of the group have a single component of single-stranded RNA of molecular weight 2.0×10^6. Some isolates of BWYV have two RNA species. The viruses have a single major protein of molecular weight about 24,000. The viruses are confined to the phloem tissues of infected plants, where they induce degeneration. All luteoviruses thus far described induce interveinal yellowing and/or the production of yellow vein on infected plants. They are all transmitted by aphids in a persistent manner and are not mechanically transmitted.

Three whitefly-transmitted viruses that are persistent in their whitefly vectors, *Bhendi* yellow mosaic, *Eclipta* yellow vein mosaic, and *Zinnia* yellow net, are similar in that they induce interveinal yellowing and yellow vein symptoms and have not been mechanically transmitted (Table 3.6). They seem to be distinct from the rod-shaped viruses in their persistence in the vector and from the geminiviruses in regard to symptom expression. Although no particles have been visualized from these viruses, they induce symptoms and have vector relationships not unlike the luteoviruses.

Rod-Shaped DNA Viruses

A semipersistent rod-shaped virus with particles 740–800 × 15–18 nm has been reported from Israel (Table 3.7). The virus, cucumber vein yellowing virus (CVYV), consists of double-stranded DNA and has a single protein subunit of 3.9×10^4 molecular weight (23). It is unique among plant viruses, since other DNA-containing plant viruses have spherical symmetry.

TABLE 3.5. Whitefly-transmitted virus with possible relationship to the potyviruses.

Virus	Particles observed	Serological relationship	Vector	Type of persistence	Transmission through molt	Symptoms[a]	Mechanical transmission	Reference
Sweet potato mild mottle	Rod 800–950 nm	—	Bemisia	—	—	M	Yes	53

[b]Type of symptoms: M, mottle.

TABLE 3.6. Whitefly-transmitted viruses with possible relationships to the luteoviruses.

Virus	Particles observed	Serological relationship	Vector	Type of persistence[a]	Transmission through molt	Symptoms[b]	Mechanical transmission	Reference
Bhendi yellow vein mosaic	—	—	Bemisia	P	—	Y, VC	—	88, 89
Eclipta yellow vein mosaic	—	—	Bemisia	P	—	Y, YV	No	10
Zinnia yellow net	—	—	Bemisia	P	—	Y, YV	No	78

[a]Type of persistence: P, persistent.
[b]Type of symptoms: VC, vein clearing; Y, interveinal yellowing; YV, yellow vein.

TABLE 3.7. Whitefly-transmitted DNA-containing rod shaped virus.

Virus	Particles observed	Serological relationship	Vector	Type of persistence[a]	Transmission through molt	Symptoms[b]	Mechanical transmission	Reference
Cucumber vein yellowing	Rod 740–900 DNA	—	Bemisia	SP	—	YV	Yes	26, 69, 75

[a]Type of persistence: SP, semipersistent.
[b]Type of symptoms: YV, yellow vein.

TABLE 3.8. Whitefly-transmitted virus with possible relationship to the nepoviruses.

Virus	Particles observed	Serological relationship	Vector	Type of persistence[a]	Transmission through molt	Symptoms[b]	Mechanical transmission	Reference
Tomato necrotic dwarf	Isometric 30 nm	—	Bemisia	NP	—	M, RS	Yes	59

[a]Type of persistence: NP, nonpersistent.
[b]Type of symptoms: M, mottle; RS, ringspot.

TABLE 3.9. Whitefly-transmitted viruses with unknown relationships.

Virus	Particles observed	Serological relationship	Vector	Type of persistence[a]	Transmission through molt	Symptoms[b]	Mechanical transmission	Reference
Anthurium mosaic	—	—	Bemisia	—	—	M	—	51
Blumea yellow vein mosaic	—	—	Bemisia	—	—	M, YV	—	100
Cotton mosaic	—	—	Bemisia	—	—	M	—	7, 61
Eupatorium pseudo mosaic	—	—	Bemisia	—	—	M	—	85
Lagendra yellow vein mosaic	—	—	Bemisia	—	—	YV	—	90
Leucas yellow vein mosaic	—	—	Bemisia	—	—	YV	No	90
Pumpkin yellow vein mosaic	—	—	Bemisia	SP	—	YV	No	89, 90
Rose yellow mosaic	—	—	Bemisia	—	—	YV	—	6
Salvia yellow vein mosaic	—	—	Bemisia	—	—	YV	No	93
Tobacco yellow net	—	—	Bemisia	—	—	YV	No	39

[a]Type of persistence: SP, semipersistent.
[b]Type of symptoms: M, mottle; YV, yellow vein.

Viruses with Possible Relationships to the Nepoviruses

A newly discovered whitefly-transmitted virus, tomato necrotic dwarf (TNDV), which affects tomato, pepper, and other Solanaceous crops and weeds, has been shown to contain three distinct isometric components ~30 nm in diameter (59) (Table 3.8). The middle and bottom components contain single-stranded RNA of 1.8×10^6 and 2.7×10^6 molecular weight, respectively. Sedimentation coefficients of the three types of particle were 57 S, 117 S, and 138 S. Mechanical inoculation tests indicated that both middle and bottom components are required for infection.

Viruses with apparently close affinities to TNDV are the nepoviruses and the comoviruses. These two groups have bipartite genomes, isometric particles ~30 nm in diameter, and three sedimentary components. The nepoviruses are transmitted by nematodes and the comoviruses are transmitted by beetles. TNDV appears to have two proteins, a feature shared by the comoviruses and at least one tentative nepovirus.

Whitefly-Transmitted Viruses with Unknown Relationships

A group of ten or so viruses have so little information in regard to mode of transmission, particle type, etc., that no attempt to categorize them was made (Table 3.9).

It is apparent when reviewing the literature that the whitefly-transmitted diseases are an attractive challenge. The viruses are much more diverse than originally thought in regard to mode of transmission and particle type. Studies on transmission, persistence, and transmission through the molt would aid in the characterization of these viruses and may aid in their control through cultural practice modification.

References

1. Ahmed, M., 1978, Whitefly *(Bemisia tabaci)* transmission of a yellow mosaic disease of cowpea, *Vigna unguiculata, Plant Dis. Rep.* **62:**224–226.
2. Allard, H.A., 1916, A specific mosaic disease in *Nicotiana viscosa* distinct from mosaic of tobacco, *J. Agric. Res.* **7:**481–486.
3. Allen, R.M., Tucker, H., and Nelson, R.A., 1960, Leaf crumple disease of cotton in Arizona, *Plant Dis. Rep.* **44:**246–250.
4. Andrews, F.W., 1936, The effect of leaf curl disease on the yield of the cotton plant, *Emp. Cotton Grow. Rev.* **13:**287–293.
5. Benigno, D.A., Fauali-Hedayat, M.A., and Retuerma, M.L., 1975, Sampagnita yellow ring spot mosaic, *Phillip. Phytopathol.* **11:**91–92.
6. Bhargava, K.S., and Joshi, R.D., 1962, Yellow mosaic a virus disease of rose in Gorakhpur, *Sci. Cult.* **28:**184–185.

7. Bink, F.A., 1973, A new contribution to the study of cotton mosaic in Chad. I. Symptoms, transmission by *Bemisia tabaci* Genn. II. Observations on *B. tabaci*. III. Other virus diseases on cotton and related plants. *Cotton Fibres Trop*. **28:**365–378.
8. Bird, J., 1957, A whitefly transmitted mosaic of *Jatropha gossypifolia, Tech. Pap. Agric. Exp. Stn.* **22:**35.
9. Bird, J., 1962, A whitefly-transmitted mosaic of *Rhychosia minima* and its relation to tobacco leaf curl and other virus diseases of plants in Puerto Rico, *Phytopathology* **52:**285–288.
10. Bird, J., and Maramorosch, K., 1978, Viruses and diseases associated with whiteflies, *Adv. Virus Res.* **22:**55–109.
11. Bird, J., and Sanchez, J., 1971, Whitefly-transmitted viruses in Puerto Rico, *J. Agric. Univ. P. Rico* **55:**461–467.
12. Bird, J., Perez, J.E., Alconero, R., Vakili, N.C., and Melendez, P.L., 1972, A whitefly transmitted golden yellow mosaic virus of *Phaseolus lunatus* in Puerto Rico, *J. Agric. Univ. P. Rico* **56:**64–74.
13. Bird, J., Sanchez, J., Rodiguez, R.L., and Julia, F.J., 1975, Rugaceous (whitefly-transmitted) viruses in Puerto Rico, in: J. Bird and K. Maramorosch (eds.), *Tropical Diseases of Legumes,* Academic Press, New York, pp. 3–25.
14. Bisht, N.S., and Mathur, R.S., 1964, Occurrence of two strains of Jute mosaic virus in U.P., *Curr. Sci.* **33:**434–435.
15. Brown, J.K., and Nelson, M.R., 1984, Geminate particles associated with cotton leaf crumple disease in Arizona, *Phytopathology* **74:**987–990.
16. Brunt, A.A., and Kenten, R.H., 1973, Cowpea mild mottle, a newly recognized virus infecting cowpea *(Vigna unguiculata)* in Ghana, *Ann. Appl. Biol.* **74:**67–74.
17. Chant, S.R., 1958, Studies on the transmission of cassava mosaic virus by *Bemisia* spp. (Aleyrodidae), *Ann. Appl. Biol.* **46:**210–215.
18. Chenulu, V.V., and Phatak, H.C., 1965, Yellow mosaic of *Acalypha indica* L., a new whitefly transmitted disease from India, *Curr. Sci.* **34:**321–322.
19. Clerk, G.C., 1960, A vein-clearing virus of sweet potato in Ghana, *Plant Dis. Rep.* **44:**931–933.
20. Cohen, S., 1967, The occurrence in the body of *Bemisia tabaci* of a factor apparently related to the phenomenon of "Periodic acquisition of tomato yellow leaf curl virus," *Virology* **31:**180–183.
21. Cohen, S., 1969, *In vivo* effects in whiteflies of a possible antiviral factor, *Virology* **37:**448–454.
22. Cohen, S., 1982, Control of whitefly vectors of viruses by colour mulches, in: K.F. Harris and K. Maramorosch (eds.), *Pathogen, Vectors, and Plant Diseases, Approaches to Control,* Academic Press, New York, pp. 45–56.
23. Cohen, S., and Antignus, Y., 1982, A noncirculative whitefly-borne virus affecting tomatoes in Israel, *Phytoparasitica* **10:**101–109.
24. Cohen, S., and Harpaz, I., 1964, Periodic, rather than continual acquisition of a new tomato virus by its vector, the tobacco whitefly (*Bemisia tabaci* Genn.), *Entomol. Exp. appl.* **7:**155–166.
25. Cohen, S., and Marco, S., 1970, Periodic occurrence of an anti-TMV factor in the body of whiteflies carrying the tomato yellow leaf curl virus (TYLCV), *Virology* **40:**363–368.

26. Cohen, S., and Nitzany, F.E., 1960, A whitefly-transmitted virus of cucurbits in Israel, *Phytopathol. Medit.* **1:**44–46.
27. Cohen, S., and Nitzany, F.E., 1966, Transmission and host range of the tomato yellow leaf curl virus, *Phytopathology* **56:**1127–1131.
28. Cohen, S., Melamed-Makjar, V., and Hameiri, J., 1974, Prevention of the spread of tomato yellow leaf curl virus transmitted by *Bemisia tabaci* Genn. (Homoptera, Aleyrodiadae) in Israel, *Bull. Entomol. Res.* **64:**193–197.
29. Cohen, S., Duffus, J.E., Larsen, R.C., Liu, H.Y., and Flock, R.A., 1983, Purification, serology, and vector relationships of squash leaf curl virus, a whitefly-transmitted geminivirus, *Phytopathology* **73:**1669–1673.
30. Costa, A.S., 1965, Three whitefly-transmitted virus diseases of beans in Sao Paulo, Brazil, *Plant Protect. Bull. FAO* **13:**121–130.
31. Costa, A.S., 1969, Whiteflies as virus vectors, in: K. Maramorosch (ed.), *Viruses, Vectors, and Vegetation,* Wiley, New York, pp. 95–119.
32. Costa, A.S., 1975, Increase in the populational density of *Bemisia tabaci,* a threat of widespread virus infection of legume crops in Brazil, in: J. Bird and K. Maramorosch (eds.), *Tropical Disease of Legumes,* Academic Press, New York, pp. 27–49.
33. Costa, A.S., 1976, Whitefly-transmitted diseases, *Annu. Rev. Phytopathol.* **14:**429–449.
34. Costa, A.S., and Bennett, C.W., 1950, Whitefly-transmitted mosaic of *Euphorbia prunifolia, Phytopathology* **40:**266–283.
35. Costa, A.S., and Carvalho, A.M.B., 1960a, Mechanical transmission of the *Abutilon* mosaic virus, *Phytopathol. Z.* **37:**259–272.
36. Costa, A.S., and Carvalho, A.M.B., 1960b, Comparative studies between *Abutilon* and *Euphorbia* mosaic virus, *Phytopathol. Z.* **38:**129–152.
37. Debrot, C.E., Herold, F., and Dao, F., 1963, Preliminary note on yellowish mosaic of tomato in Venezuela, *Agron. Trop. Maracay* **13:**33–41.
38. Dhanraj, K.S., and Seth, M.L., 1968, Enations in *Capsicum annum* (Chilli) caused by a new strain of leaf curl virus, *Ind. J. Hortic.* **25:**70–71.
39. Dhingra, K.L., and Nariani, T.K., 1961, Yellow net virus disease of tobacco plant, *Ind. J. Microbiol.* **1:**94–99.
40. Dickson, R.C., Johnson, M.McD., and Laird, E.F., 1954, Leaf crumple, a virus disease of cotton, *Phytopathology* **44:**479–480.
41. Duffus, J.E., 1965, Beet pseudo-yellows virus, transmitted by the greenhouse whitefly *(Trialeurodes vaporariorum), Phytopathology* **55:**450–453.
42. Duffus, J.E., 1975, A new type of whitefly-transmitted disease. A link to the aphid-transmitted viruses, in: J. Bird and K. Maramorosch (eds.), *Tropical Diseases of Legumes,* Academic Press, New York, pp. 79–88.
43. Duffus, J.E., and Flock, R.A., 1982, Whitefly-transmitted disease complex of the desert southwest, *Calif. Agric.* **36:**4–6.
44. Duffus, J.E., and Johnstone, G.R., 1981, Beet pseudo yellows virus in Tasmania. The first report of a whitefly transmitted virus in Australasia, *Australasian Plant Pathol.* **10:**68–69.
45. Duffus, J.E., Mayhew, D.E., and Flock, R.A., 1982, Lettuce infectious yellows—a new whitefly transmitted virus of the desert southwest, *Phytopathology* **72:**963.
46. Duffus, J.E., Larsen, R.C., and Liu, H.Y., 1986, Lettuce infectious yellows virus—A new type of whitefly transmitted virus, *Phytopathology* **76:**97–100.

47. Duffus, J.E., Liu, H.-Y., and Johns, M.R., 1985, Melon leaf curl virus—A new geminivirus with host and serological variations from squash leaf curl virus, *Phytopathology* **75:**1312.
48. Gadd, C.H., and Loos, C.A., 1941, A virus disease of *Ageratum conyzoides* and tobacco, *Trop. Agric.* **96:**255–264.
49. Goodman, R.M., 1977, Single-stranded DNA genome in a whitefly-transmitted plant virus, *Virology* **83:**171–179.
50. Goodman, R.M., Bird, J., and Thongmeearkom, P., 1977, An unusual virus-like particle associated with golden yellow mosaic of beans, *Phytopathology* **67:**37–42.
51. Herold, F., 1967, Investigation of a virus disease of *Anthorium andraenum, Phytopathology* **57:**8.
52. Hoefert, L.L., 1983, Ultrastructure of *Cucurbita* spp. infected with whitefly-transmitted squash leaf curl virus, *Phytopathology* **75:**790.
53. Hollings, M., Stone, O.M., and Bock, K.R., 1976, Purification and properties of sweet potato mild mottle, a whitefly-borne virus from sweet potato *(Ipomoea batatas)* in East Africa, *Ann. Appl. Biol.* **82:**511–528.
54. Iwaki, M., Thongmeearkom, P., Prommin, M., Honda, Y., and Hibi, T., 1982, Whitefly transmission and some properties of cowpea mild mottle virus on soybean in Thailand, *Plant Dis.* **66:**365–68.
55. Kim, K.S., and Martin, E.M., 1982, Nucleolar and extranucleolar perichromatin granules induced by *Euphorbia* mosaic virus, *Phytopathology* **72:**938.
56. Kim, K.S., Shock, T.L., and Goodman, R.M., 1978, Infection of *Phaseolus vulgaris* by bean golden mosaic virus: Ultrastructural aspects, *Virology* **89:**22–33.
57. Kirkpatrick, T.W., 1930, Leaf curl in cotton, *Nature* **125:**85–97.
58. Larsen, R.C., and Kim, K.S., 1985, Ultrastructure of *Diodia virginiana* infected with a whitefly-transmitted virus-like disease agent, *Phytopathology.* **75:**1324.
59. Larsen, R.C., Duffus, J.E., and Liu, H.Y., 1984, Tomato necrotic dwarf a new type of whitefly-transmitted virus, *Phytopathology* **74:**795.
60. Lot, H., Onillon, J.C., and Lecoq, H., 1980, Une nouvelle maladie a virus de la laitue de serre: La jaunisse transmise par la mouche blanche, *Rev. Hortic.* **209:**31–34.
61. Lourens, J.H., Laon, P.A.V., and Brader, L., 1972, Contribution to the study of a cotton mosaic in Chad: Distribution in a field: Common Aleurodidae; transmission trials from cotton to cotton by Aleurodidae, *Cotton Fibres Trop.* **27:**225–230.
62. Mathur, R.N., 1933, Leaf curl in *Zinnia elegans* at Dehra Dun, *Ind. J. Agric. Sci.* **3:**89–96.
63. Muniyappa, V., 1980, Whiteflies, in: K. Harris and K. Maramorosch (eds.), *Vectors of Plant Pathogens,* Academic Press, New York, pp. 39–85.
64. Muniyappa, V., and Reddy, D.V.R., 1983, Transmission of cowpea mild mottle virus by *Bemisia tabaci* in a nonpersistent manner, *Plant Dis.* **67:**391–393.
65. Nair, R.R., and Wilson, K.I., 1969, Studies on some whitefly-transmitted plant virus diseases from Kerala, *Agric. Res. J. Kerala* **7:**123–126.
66. Nair, R.R., and Wilson, K.I., 1970, Leaf curl of *Jatropha curcas* L. in Kerala, *Sci. Cult.* **36:**569.

67. Nariani, T.K., and Pathanian, P.S., 1953, *Physalis peruviana* L. a new host of tobacco leaf curl virus, *Ind. Phytopathol.* **6:**143–145.
68. Nene, Y.L., 1972, A survey of viral diseases of pulse crops in UP, *G.B. Plant Univ. Agric. Tech.* Bull. No. 4 191 pp.
69. Nitzany, F.E., Geisenberg, H., and Koch, B., 1964, Tests for the protection of cucumbers from a whitefly-borne virus, *Phytopathology* **54:**1059–1061.
70. Orlando, A., and Silberschmidt, K., 1946, Studies on the natural dissemination of the virus of infectious chlorosis of the Malvaceae (*Abutilon* virus 1. Baur) and its relation with the insect *Bemisia tabaci* (Homoptera-Aleyrodidae), *Arq. Inst. Biol. S. Paulo* **17:**1–36.
71. Ramakrishnan, K., Kandaswamy, T.K., Subramanian, K.S., Janarthanan, R., Mariappan, V., Samuel, G.S., and Navaneethan, G., 1971, Investigations on virus diseases of pulse crops in Tamil Nadu, Final Technical Report, Tamil Nadu Agricultural University, Coimbatore, India.
72. Roberts, I.M., Robinson, D.J., and Harrison, B.D., 1984, Serological relationships and genome homologies among geminiviruses, *J. Gen. Virol.* **65:**1723–1730.
73. Sahambi, H.S., 1958, in: *Proceedings of the Mycological Research Workers Conference,* Simla, pp. 181–84.
74. Schuster, M.F., 1964, A whitefly-transmitted mosaic virus of *Wissadula amplissima, Plant Dis. Rep.* **48:**902–905.
75. Sela, I., Assouline, I., Tanne, E., Cohen, S., and Marco, S., 1980, Isolation and characterization of rod-shaped, whitefly-transmissible, DNA-containing plant virus, *Phytopathology* **70:**226–228.
76. Singh, B.P., and Misra, A.K., 1971, Occurrence of hollyhock yellow mosaic virus in India, *Ind. Phytopathol.* **24:**213–214.
77. Smith, K.M., 1957, Hollyhock mosaic virus, in: K.M. Smith (ed.), *Plant Virus Diseases,* Churchill, London. 641 pp.
78. Srivastava, K.M., Singh, B.P., Dwadash Shreni, V.C., and Srivastava, B.N., 1977, *Zinnia* yellow net disease—Transmission, host range, and agent–vector relationship, *Plant Dis. Rep.* **61:**550–554.
79. Stein, V.E., Coutts, R.H.A., and Buck, K.W., 1983, Serological studies on tomato golden mosaic virus, a geminivirus, *J. Gen. Virol.* **64:**2493–2498.
80. Storey, H.H., 1931, A new virus disease of the tobacco plant, *Nature* **128:**187–188.
81. Storey, H.H., and Nichols, R.F.W., 1938, Studies of the mosaic of cassava, *Ann. Appl. Biol.* **25:**790–806.
82. Sylvester, E.S., 1956, Aphid transmission of plant viruses, in: *Proceedings of the 10th International Congress of Entomology* Vol. 3, pp. 195–200.
83. Takami, N., 1901, On stunt disease of rice plant and *Nephotettix apicalis* Motsch. var. Cincticeps Uhl, *J. Jpn. Agric. Soc.* **241:**22–30.
84. Tarr, S.A.J., 1951, Leaf curl disease of cotton, Commonwealth Mycological Institute, Kew, Surrey, England, 55 pp.
85. Van Der Laan, P.A., 1940, Motschilduis en *Eupatorium* als Oorzaken Van Pseudo-mosaic (whitefly and *Eupatorium* as causes of pseudomosaic), *Vlugschr. Deli-Proefst, Medan* **67:**4.
86. Van Dorst, H.J.M., Huijberts, N., and Bos, L., 1980, A whitefly transmitted disease of glasshouse vegetables, a novelty for Europe, *Neth. J. Plant Pathol.* **85:**311–313.

87. Van Schaik, P.H., Erwin, D.C., and Garber, M.J., 1962, Effects of time of symptom expression of the leaf crumple virus on yield and quality of fiber of cotton, *Crop Sci.* **2**:275–277.
88. Varma, P.M., 1952, Studies on the relationship of the Bhendi yellow-vein mosaic virus and its vector, the whitefly *(Bemisia tabaci), Ind. J. Agric. Sci.* **22**:75–91.
89. Varma, P.M., 1955, Persistance of yellow-vein virus of *Abelmoschus esculentus* (L.) Moench in its vector *Bemisia tabaci* (Gen.), *Ind. J. Agric. Sci.* **25**:293–302.
90. Varma, P.M., 1963, Transmission of plant viruses by whiteflies, *Bull. Nat. Inst. Sci. India* **24**:11–33.
91. Varma, P.M., Rao, D.G., and Capoor, S.P., 1966, Yellow mosaic of *Corchorus trilocularis, Sci. Cult.* **32**:466.
92. Verma, H.N., Srivastava, K.M., and Mathur, A.K., 1975, A whitefly-transmitted yellow mosaic virus disease of tomato from India, *Plant Dis. Rep.* **59**:494–498.
93. Verma, V.S., 1974a, *Salvia* yellow-vein mosaic virus, *Gartenbauwissenschaft* **39**:565–566.
94. Verma, V.S., 1974b, Lupin leaf curl virus, *Gartenbauwissenschaft* **39**:55–56.
95. Verma, V.S., 1974c, Soapwort leaf curl virus, *Gartenbauwissenschaft* **39**:567–568.
96. Verma, V.S., and Singh, S., 1973a, Balsam leaf curl disease, *Hortic. Res.* **13**:55–56.
97. Verma, V.S., and Singh, S., 1973b, *Zinnia* leaf curl virus, *Gartenbauwissenschaft* **38**:159–162.
98. Vasudeva, R.S., Varma, P.M., and Capoor, S.P., 1953, Some important whitefly- (*Bemisia tabaci* Gen.) transmitted virus in India, *Atti VI International Congress of Microbiology Roma,* Vol. 5, pp. 520–521.
99. Wilson, K.I., 1972, Chlorotic ring spot of Jasmine, *Ind. Phytopathol.* **25**:157–158.
100. Wilson, K.I., and Potty, V.P., 1972, Yellow vein mosaic of *Blumea neilgherrensis, Agric. Res. J. Kerala* **10**:68.
101. Wolf, F.A., Whitecomb, W.H., and Mooney, W.C., 1949, Leaf curl of tobacco in Venezuela, *J. Elisha Mitchell Sci. Soc.* **65**:38–47.
102. Yamashita, S., Doi, Y., Yora, K., and Yoshino, M., 1979, Cucumber yellows virus: Its transmission by the greenhouse whitefly, *Trialeurodes vaporariorum* (Westwood), and the yellowing disease of cucumber and muskmelon caused by the virus, *Ann. Phytopathol. Soc. Japan* **45**:484–496.

4
Virus–Membrane Interactions Involved in Circulative Transmission of Luteoviruses by Aphids

Frederick E. Gildow

Introduction

Aphids are ideally suited for transferring plant-infecting viruses from host to host. The piercing–sucking mouthparts, mode of dispersal, and prolific reproductive behavior of aphids facilitate survival and spread of plant viruses over large areas. Two general mechanisms have been described to explain how aphids transmit plant viruses. One mechanism involves aphid acquisition of virus during brief probes of the aphid's stylet into epidermal cells of virus-infected plants. During these exploratory test probes the stylets penetrate epidermal cells and the aphid ingests cell contents into the fore gut, especially the maxillary food canal, cibarial pump, and pharynx. During ingestion some viruses suspended in the plant cytoplasm are believed to adhere to the lining of the fore gut (27, 28, 55). Adherence of virus to the chitin-lined fore gut may involve an interaction between the virus and other compounds (helper factor) produced in the infected cell (4). If the aphid completes a test probe on an infected plant and then moves immediately to another plant and initiates a second test probe, it is believed that some of the virions in the lumen are adhering to the lining of the fore gut may detach and be expelled or egested into the newly penetrated epidermal cell being sampled (27, 28). For unknown reasons aphids lose the ability to transmit viruses by this mechanism within minutes or hours. This type of transmission is, therefore, described as non-persistent transmission. Other types of viruses are probably transmitted by a similar mechanism; however, aphids retain the ability to transmit these viruses for up to several days. This type of transmission has been described as semipersistent (27, 28, 46).

The focus of this chapter will be on a second general mechanism for plant virus transmission by aphids involving viruses that are retained and

---

Frederick E. Gildow, Department of Plant Pathology, Pennsylvania State University, University Park, Pennsylvania 16802, USA.

transmitted by aphids over a much longer time period. These viruses are usually acquired during the longer duration feeding probes of the stylets into phloem tissues of the plant vascular system. Once the aphid has acquired this type of virus it can transmit the virus for many days or weeks. This type of transmission is referred to as persistent transmission. Unlike the rather superficial association between the virus and aphid resulting in nonpersistent transmission, persistently transmitted viruses have a very intimate association with their vectors. To be transmitted, these viruses must be actively acquired and subsequently released by vector cells, while remaining infectious to the plant host. This type of transmission is also called circulative (28) because these viruses circulate throughout the body of their aphid vectors (Fig. 4.1). Most aphid-transmitted circulative viruses do not replicate in their aphid vectors (7, 18, 30, 50, 71). Evidence does indicate that at least one group of viruses, the rhabdoviruses, contains member viruses that infect and replicate in both plant host and aphid vector tissues. The biology of these propagative, circulatively transmitted viruses has been reviewed (28, 69).

The ability of a plant virus to be aphid-transmitted, and whether the

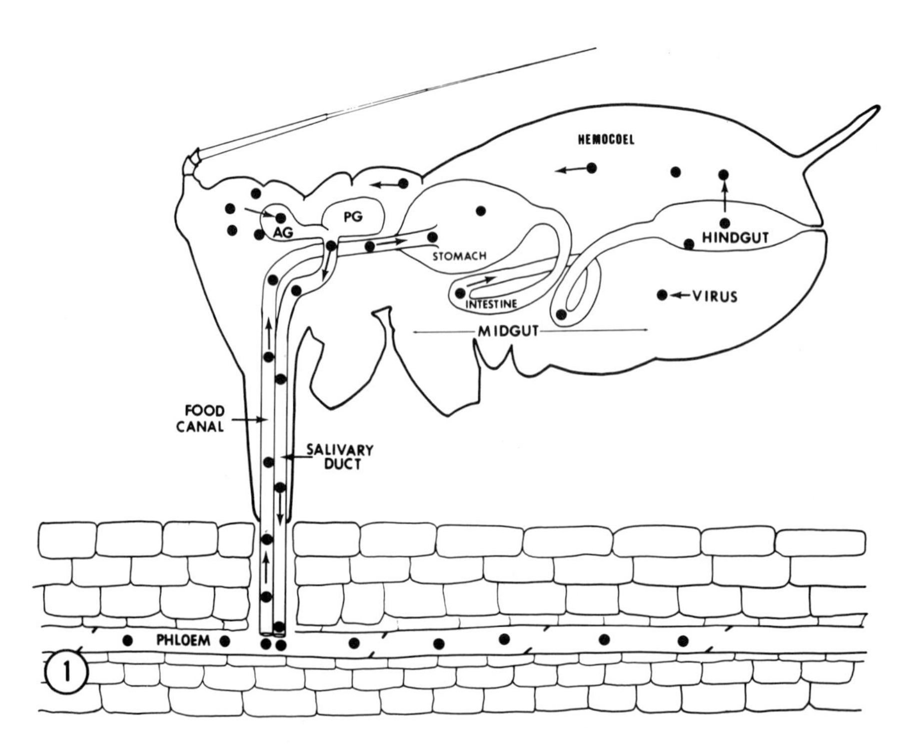

FIGURE 4.1. Diagram of the route of circulatively transmitted luteoviruses through the aphid vector; accessory salivary gland (AG), principal salivary gland (PG).

virus is transmitted in a nonpersistent or persistent manner, are characteristics determined in part by the genetic constitution of the virus (31). Host tissue specificity of the virus, the virion capsid protein structure (2, 20, 45, 58), and the synthesis of helper factors in infected cells (4) are some of the characteristics determined by the virus genome that seem to play key roles in regulating virus transmission mechanisms. The genetic traits of the vector must also play an important role in directing virus transmission, but are less well understood.

Aphid and virus characteristics related to circulative virus transmission have been described in a general way for several aphid–virus systems (8, 14–16, 28, 42, 48, 57). The purpose of this chapter, therefore, will be to describe in detail more recent work directed at better understanding the cellular mechanisms that regulate plant virus recognition and transport through aphid vector tissues. The virus–membrane interactions involved in vector-specific transmission of luteoviruses will be described as a model system for studying the complex of interactions regulating circulative virus transmission.

Characteristics of Luteoviruses

Plant viruses in the luteovirus group (60) have small icosahedral virions, approximately 25 nm in diameter, consisting of a single-stranded RNA genome (MW = 2×10^6 daltons), which is encapsidated in a naked protein shell constructed of a single major coat protein subunit (2.4×10^4 daltons). The most studied members of the luteovirus group are barley yellow dwarf virus (BYDV), beet western yellows virus (BWYV), and potato leaf roll virus (PLRV). Although host ranges of the various luteoviruses encompass many species of monocot and dicot plants, symptoms induced in plants following infection by different luteoviruses are similar. This is probably due to the fact that luteoviruses are host tissue-specific and replicate primarily in phloem cells (37, 67). Infection leads to phloem necrosis, which disrupts the normal translocation pathways of plant metabolites (38). As a result, plants infected with luteoviruses frequently exhibit reduced growth and chlorotic yellowed foliage.

Luteoviruses are transmitted only by aphids and have not been mechanically inoculated to plants. This may be due to the phloem specificity of the viruses. One of the most interesting characteristics of luteoviruses is their vector specificity. Each luteovirus is transmitted efficiently by only one or a few species of aphids. The virus–vector relationships of BYDV have been the most intensively studied. Five characterized, serologically distinct (59) isolates of BYDV were first identified on the basis of their aphid transmission specificity (39, 57). The most efficient vectors of each of these isolates is shown in Table 4.1. The cellular and molecular basis determining this specificity is the topic of this chapter.

TABLE 4.1. Transmission patterns of cereal grain aphid species for five characterized New York isolates of barley yellow dwarf virus (BYDV).

	BYDV isolates[a]				
Aphid species[b]	RPV	RMV	MAV	PAV	SGV
Rhopalosiphum padi	++	−	−	++	−
Rhopalosiphum maidis	−	++	−	−	−
Metopolophium dirhodum	−	−	++	+	−
Sitobion avenae	−	−	++	++	−
Schizaphis graminum	+	−	−	+	++

[a] ++, an efficient vector with a greater than 50% probability of transmission in single aphid tests; +, an inefficient vector with less than 50% probability of transmission; −, a nonvector with less than 5% probability of transmission by single aphids.
[b] New York aphid clones maintained at Cornell University by W.F. Rochow, except for *M. dirhodum* collected at Berkeley, California.

The Route of Luteoviruses Through Aphid Vectors

In order for luteovirus to be acquired for transmission, the aphid vector must initiate a feeding probe into the infected plant host. During a feeding probe, stylets are inserted intercellularly through the epidermis and underlying tissue (cortex or mesophyl) until they reach a vascular bundle (Fig. 4.2). Evidence indicates that aphid salivary secretions contain enzymes with pectinase and cellulase activity, which aid stylet penetration through the middle lamella, a pectin-containing layer formed between plant cells (44). Once the stylets reach the phloem tissue within the vascular bundle, the stylets preferentially penetrate a sieve tube element. The aphid begins feeding by activating large muscles in the head, which enlarge the lumen of the pharynx, creating a reduced pressure in the food canal. Contents of the penetrated plant sieve tube element are sucked into the food canal.

Luteovirus ingestion occurs when the aphid feeds on infected phloem cells (Fig. 4.1). Ingested virus moves through the food canal to the foregut. Unlike the nonpersistently transmitted plant viruses, attachment of persistently transmitted viruses to the foregut is not believed to play a role in virus transmission. Instead, virus continues to move through the lumen of the enlarged stomach region of the anterior midgut and the narrower intestinal region of the posterior midgut. At the hindgut of the alimentary canal the virus may pass into the aphid body cavity or hemocoel. If this occurs, then the aphid has acquired the virus. For the purpose of this discussion, virus acquisition will imply movement of virus from the gut lumen to the hemocoel. Unacquired viruses continue to pass through the alimentary canal to the rectum and are eliminated in the voided honeydew. Acquired viruses suspended in the hemolymph move throughout the open circulating system which bathes all the aphid's organ systems. The aphid

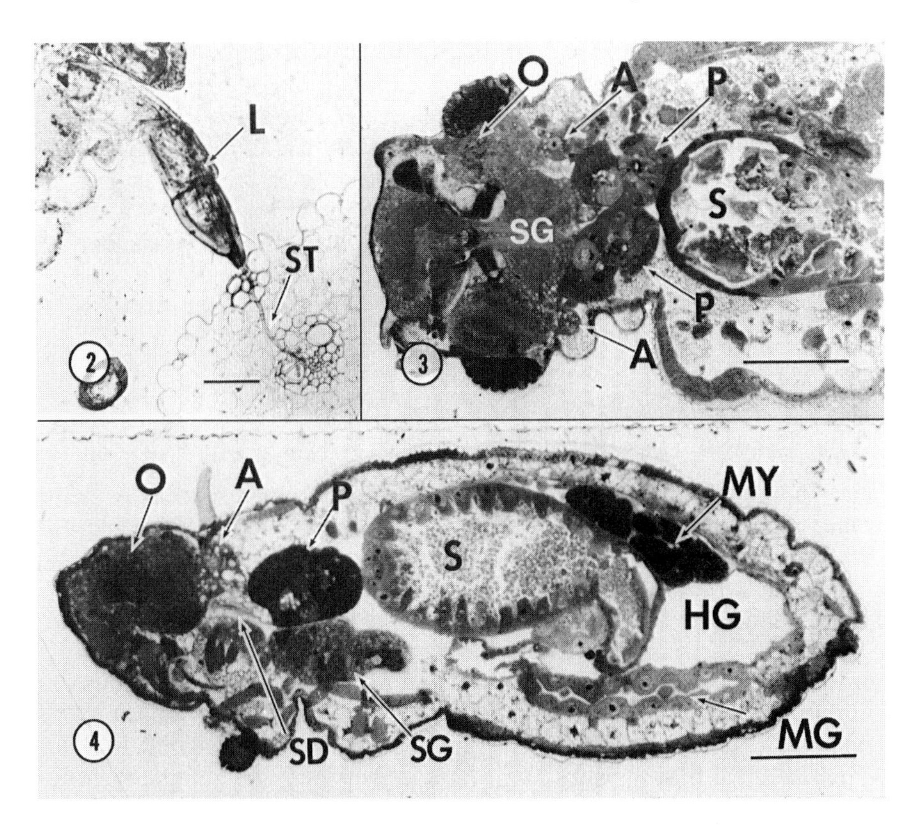

FIGURES 4.2–4.4. Light micrographs of *Rhopalosiphum padi*. FIGURE 4.2. Section through an oat leaf showing the position of the aphid labium (L) during feeding and the penetration of the stylets (ST) to the phloem tissue of a vascular bundle. FIGURE 4.3. A 0.5-μm section showing a dorsal view through the head and thorax. The position of the accessory salivary glands (A) and principal salivary glands (P), relative to the optic lobes (O), subesophageal nerve glanglion (SG), and stomach (S), are shown. FIGURE 4.4. A tangential section showing a lateral view of the anatomical relationship of the optic lobe (O), accessory salivary gland (A), principal salivary gland (P), salivary duct (SD), subesophageal nerve ganglion (SG), stomach (S) region of the anterior midgut, posterior midgut (MG), hindgut (HG), and symbiont-containing mycetocytes (MY). Bars = 0.1 mm.

becomes a reservoir for the plant virus, aiding virus dispersal and survival in the absence of suitable plant host species. To be transmitted by the aphid to another plant, the virus must pass through the accessory salivary gland and be released into the salivary duct (Figs. 4.3 and 4.4). Once this occurs the virus is anatomically external to aphid tissues and free to be excreted into plant phloem cells along with salivary gland secretory products during feeding.

Observations of Luteovirus Acquisition

The cellular site for BYDV acquisition has been recently determined (22). Ultrastructural studies of the bird cherry-oat aphid (*Rhopalosiphum padi* L.) fed on oats (*Avena byzantina* Koch) infected with the RPV isolate of BYDV indicated a specific association of RPV virions with the aphid hindgut. Although individual particles of RPV were observed in the lumen of the midgut of the alimentary canal, they were never observed in contact with midgut cell membranes. In the hindgut, however, virions were characteristically observed associated with several membrane systems in 37 of 40 viruliferous aphids examined.

Aphid Hindgut Structure

The hindgut region of the aphid alimentary canal is an extremely thin-walled structure centrally located in the abdomen (Fig. 4.5). Composed of stretched squamous epithelial cells, with areas of cytoplasm only 0.2–1.0 μm thick, the fragile nature of the hindgut makes it difficult to dissect for study and often difficult to identify in tissue sections by light microscopy (Fig. 4.6). Distinguishing features can be easily observed, however, in thin sections by transmission electron microscopy (TEM). A portion of a transverse section of hindgut, midgut, and intervening muscle cell, similar to that seen in Figure 4.6, can be seen at higher magnification in Figure 4.7. All three tissues shown are surrounded and separated from one another by a distinct basement membrane or basal lamina. The term basal lamina will be used here to avoid confusion with true membranes. The basal lamina is an extracellular fibrous network of a complex of collagen, protein, and carbohydrate (40). Basal lamina surround most major organ systems providing structural support and acting as a preliminary filtering mechanism for molecules entering or leaving the hemolymph. The most distinguishing feature of the hindgut is the single row of extracellular microtubulelike structures which line the exterior of the apical plasmalemma facing the lumen of the hindgut (49). Similar structures were not observed elsewhere in the aphid and their function is unknown. The hindgut apical and basal cell membranes are not invaginated into the cytoplasm as they are in the midgut and there are fewer microvilli formed on the hindgut surface compared to the midgut. A normal complement of subcellular organelles occurs in the hindgut cytoplasm with numerous mitochondria predominately located near the apical plasmalemma and Golgi bodies more closely associated with the basal plasmalemma adjacent to the hemocoel. Coated pits (65) are common features along the apical plasmalemma and coated vesicles (5) and rough endoplasmic reticulum are common cytoplasmic components. This cell architecture suggests that the hindgut epithelium is designed for active transport and endocytotic uptake of compounds at the apical plasmalemma.

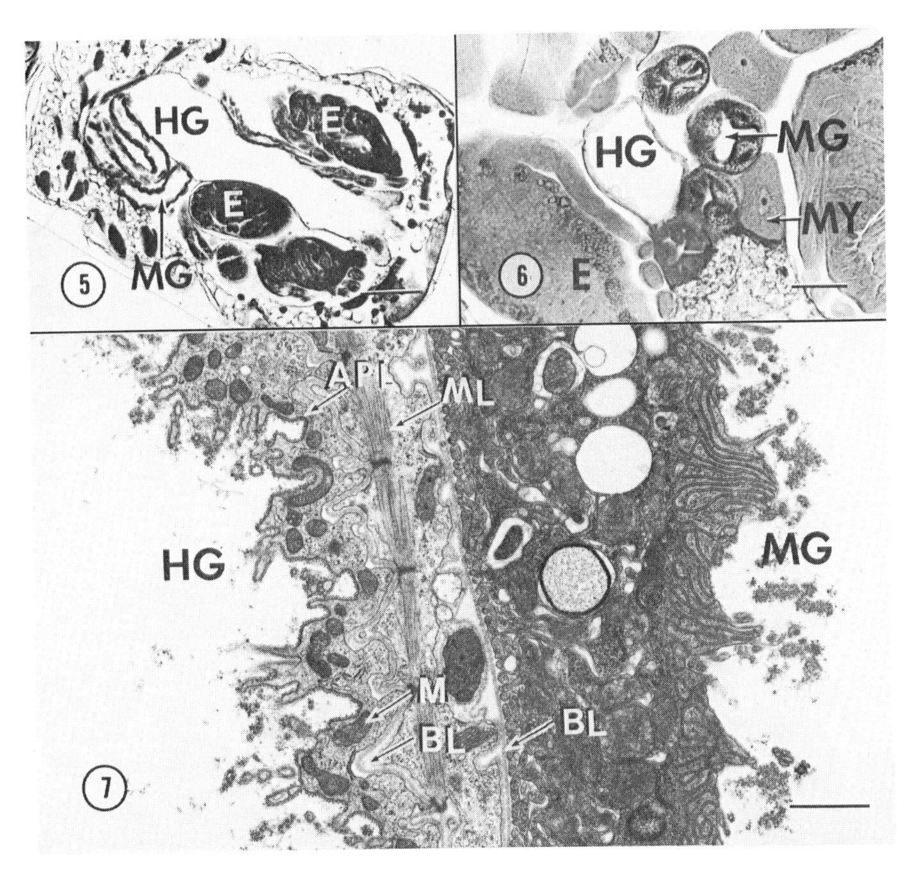

FIGURES 4.5–4.7. Midgut and hindgut structure of *Rhopalosiphum padi*. FIGURE 4.5. Light micrograph of a lateral section through the abdomen showing a dorsal view of the location and appearance of the midgut (MG) and hindgut (HG); embryos (E). Bar = 0.2 mm. FIGURE 4.6. Light micrograph of a cross section through the abdomen. Note difference in wall thickness between the hindgut (HG) and midgut (MG) sections, embryos (E), and mycetocyte (MY). Bar = 50μm. FIGURE 4.7. Electron micrograph of a transverse section through the hindgut (HG) and adjacent muscle tissue (ML) and midgut (MG). Note difference in cytoplasmic density and thickness; mitochondria (M), basal lamina (BL), apical plasmalemma (APL). Bar = 1 μm.

Barley Yellow Dwarf Virus Acquisition

In *R. padi* fed on RPV-infected oats, virus particles were consistently observed associated with the hindgut apical plasmalemma. Although virions were rarely observed freely suspended in the lumen, they were observed apparently attached to the apical cell membrane (Fig. 4.8). Most particles at the apical membrane were observed individually in shallow

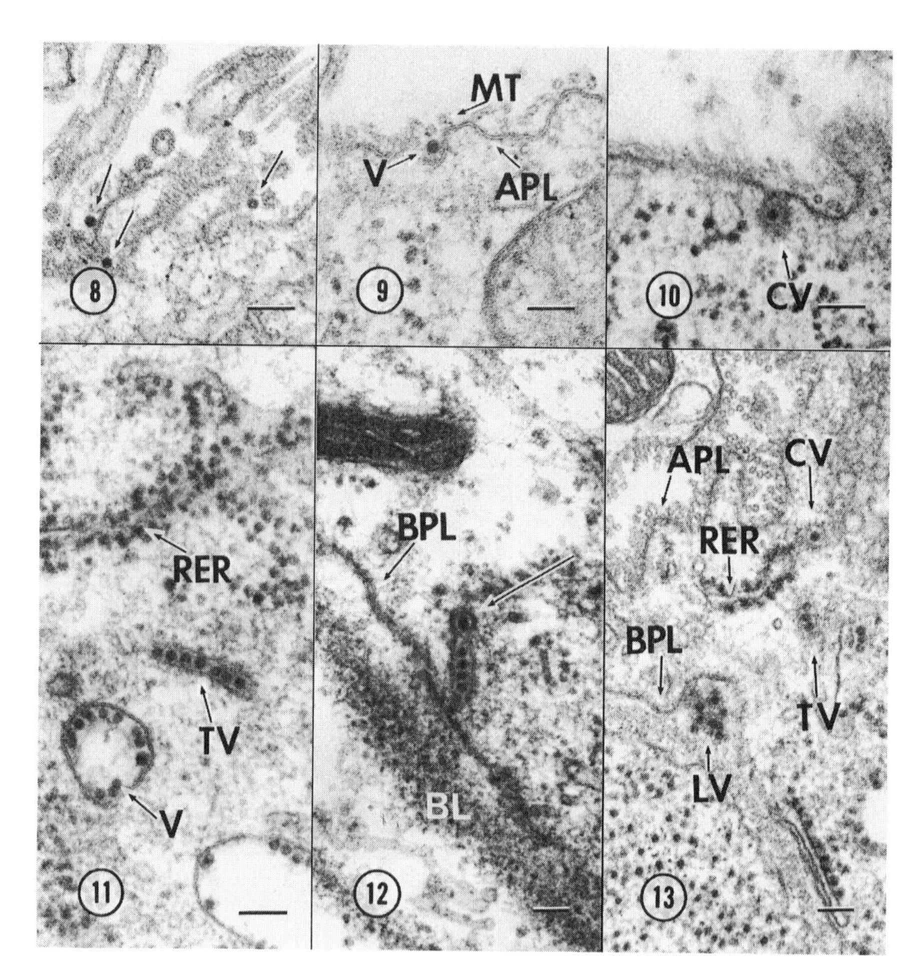

FIGURES 4.8–4.13. Electron micrographs showing the ultrastructural association of the RPV isolate of BYDV with hindgut cells of *Rhopalosiphum padi*. FIGURE 4.8. Virions (arrows) in the lumen of the hindgut adjacent to the apical plasmalemma. FIGURE 4.9. A virus particle (V) located in a pitlike depression of the apical plasmalemma (APL) located below microtubules (MT) lining the membrane. [From Gildow (22).] FIGURE 4.10. A virion in a coated vesicle (CV) in hindgut cell cytoplasm. [From Gildow (22).] FIGURE 4.11. Virions (V) in a tubular vesicle (TV) and a vesicle adjacent to rough endoplasmic reticulum (RER) coated with ribosomes. FIGURE 4.12. A tubular vesicle containing virions (arrow) in contact with the basal plasmalemma (BPL), which is surrounded by basal lamina (BL). [From Gildow (22).] FIGURE 4.13. Unlabeled virions in a tubular vesicle (TV) and a coated vesicle (CV) adjacent to rough endoplasmic reticulum (RER), and labeled virions (LV) located external to the basal plasmalemma (BPL) and aggregated by microinjected anti-RPV antiserum; apical plasmalemma (APL). Bar = 100 nm.

pits below the layer of microtubules (Fig. 4.9). Single virions were also readily observed in coated vesicles near the plasmalemma (Fig. 4.10) and throughout the cytoplasm (Fig. 4.13). Once within the hindgut epithelial cell cytoplasm, virus particles were most readily observed to occur in linear arrays within tubular vesicles or clustered in larger, spherical vesicles (Fig. 4.11). The structural associations of virions with cellular membrane systems described above were consistently observed in viruliferous aphids and suggest mechanisms for virus transport into the hindgut cell. Associations of virion-containing membrane organelles with the basal plasmalemma were rarely observed and evidence for a mechanism of virion release into the hemocoel is less certain. Although tubular vesicles containing virions were often located near the basal cell membrane, on only a few occasions were tubular vesicles observed fused to the membrane, suggesting a possible mechanism for virion release (Fig. 4.12). With standard TEM fixation procedures (21), virions of RPV were only occasionally observed embedded in the extracellular basal lamina or between the basal lamina and the hindgut membrane indicating virus release. This suggested that under normal conditions the virus particles quickly enter the hemocoel once released from the cell. If the aphids were microinjected with antiserum to the virus, virions were readily detected aggregated in clumps extracellularly on the basal plasmalemma (Fig. 4.13). Virions being released from the hindgut were apparently aggregated and trapped in this location by homologous antibodies suspended in the hemolymph. Similar clumping did not occur when heterologous antisera to other BYDV isolates or antiserum to *R. padi* virus (12) were microinjected into the aphids. These results suggested that RPV was released from the hindgut cell, not as individual particles, but in groups, and further supported the idea that virion-containing tubular vesicles fusing to the plasmalemma could be a vehicle for virus release from the hindgut. The immunological labeling also verified identity of the observed virions as RPV. No particles or virionlike structures were ever observed in hindgut membrane systems in any *R. padi* fed on healthy oats as controls.

Characteristics of Receptor-Mediated Endocytosis and Cellular Transport

The types of membrane-bound organelle systems associated with RPV in the hindgut suggest that receptor-mediated endocytosis (52) may be involved in luteovirus acquisition by aphid vectors. Receptor-mediated endocytosis is a cellular mechanism for the internalization of macromolecules through the cell plasmalemma. Endocytosis is a common process in all eukaryotic cells for selectively acquiring and concentrating molecules suspended in fluids of the extracellular environment. Some of the intensively studied pathways involve the uptake and transport of ferric ions (26), cholesterol (6), antibodies (41), and hormones (43). Initial stages in endocytosis of the above compounds are similar. The surface of the cell contains membrane-bound proteins or glycoproteins (receptors) capable of specifically recognizing and binding certain classes of molecules (ligands) which are to be internalized. Each ligand has a distinct population of receptors and is thought to be internalized at a rate independent of other nonrelated ligand–receptor complexes. The number of receptors on the cell surface for any given ligand varies with cell type and species, but generally is reported to be in the range of 10^3–10^5 receptors per cell. To initiate the endocytotic process, a ligand molecule comes into contact with one or more of its specific receptors. A ligand–receptor complex forms, binding the molecules and holding the ligand to the membrane. Internalization begins by progressive invagination of the membrane below the ligand–receptor complex. This region of the membrane is coated on its cytoplasmic face with clathrin protein (53), giving the coated region of the membrane its characteristic appearance. The coated membrane continues to invaginate, forming obvious coated pits containing ligand–receptor. The coated pits may continue to invaginate deep into the cytoplasm, remaining attached to the plasmalemma by narrow channels (75); or they may bud off to form coated vesicles freely suspended in the cytoplasm (54). Coated vesicles are distinct organelles utilized by eukaryotic cells for transport of materials through the cytoplasm from one membrane system to another (5, 65, 76). The coated pits or vesicles eventually lose their clathrin coat, forming smooth membrane vesicles. Several of these vesicles may fuse together, forming larger, noncoated vesicles referred to as receptosomes or endosomes (51). At this point endocytosis of the ligand has been accomplished and specific molecules have been internalized and concentrated by the cell. The destination and fate of the ligand then vary with the molecule in question. Many molecules, such as galactose-exposing ligands in hepatocytes, are transferred from receptosomes to lysosomes and degraded (74). Molecules such as transferrin, used for iron uptake into cells, are transported back to the plasmalemma and recycled (26). Other molecules, such as antibodies, are transported through the cell and secreted (41). The exact route and mechanism for intracellular transport vary with the ultimate destination of the ligand in question.

Many animal-infecting viruses are known to take advantage of this normal cell process to penetrate host cell membranes (11, 72). Both membrane-bound viruses such as Semliki Forest virus (33) and budded nuclear polyhedrosis virus (73), and nonmembrane-bound viruses such as reoviruses (17) and adenoviruses (19) have been found to initiate penetration of host cells by receptor-specific binding and subsequent endocytosis into coated pits, receptosomes, and lysosomes. If the cell internalizing the virus can be infected and function as a host, the viral genome is then released into the cytoplasm by a variety of mechanisms resulting in virus replication.

A Model for Luteovirus Acquisition

In order to better discuss luteovirus acquisition through the aphid hindgut epithelium, I have developed the following model (Fig. 4.14). This model is based on morphological information provided by static ultrastructural images. The limitations of such evidence are obvious and acknowledged. The only purpose of this model is to act as a reference point to build upon, alter, or discard as more information becomes available.

In this hypothetical model the first step leading to luteovirus transport through the hindgut is an attachment of a virus particle to the apical plasmalemma of the epithelium. Attachment of the virus to the membrane by luteovirus-recognizing receptor molecules embedded in the membrane could induce endocytosis of the virion into a coated pit, which either invaginates deep into the cytoplasm or buds off to become a coated vesicle. Many virus-containing coated vesicles could eventually fuse together, forming a larger, noncoated spherical vesicle (receptosome) containing many virions. Elongated tubular vesicles could arise from a budding-off process from the receptosome in a manner similar to that previously described (51), or by direct formation by repeated end-to-end fusions of coated vesicles. Some receptosomes containing virus might fuse with lysosomes, leading to the eventual degradation of those particles. The virus-

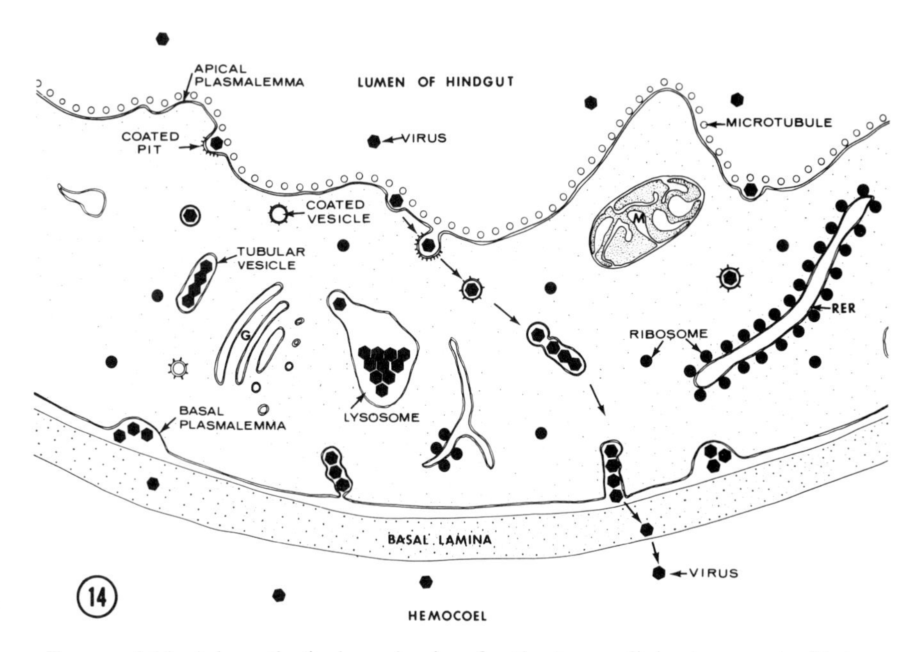

FIGURE 4.14. A hypothetical mechanism for the transcellular transport of luteoviruses through the hindgut cell cytoplasm of aphid vectors.

containing tubular vesicles, on the other hand, are directed toward the basal end of the cell and eventually contact and fuse with the basal plasmalemma, exposing the vesicle lumen, and the virus it contains, to the aphid hemocoel. The virus would then be free to diffuse through the extracellular basal lamina suspended in hemolymph. According to this model, the luteovirus is transported through the hindgut continuously packaged by a membrane system and the virus itself is never in direct contact with the cell cytoplasm.

The luteovirus model described above is somewhat analogous to the mechanism for transepithelial transfer of IgA through mammalian tissues. Although the direction of transport is the opposite of that described for luteoviruses, the ultrastructural mechanism is similar. Several immunohistochemical studies of a variety of tissues (41, 47) have demonstrated that dimeric IgA binds specifically to its receptor located on the basal plasmalemma. Results of time course studies indicated that within 1–3 min of application, IgA was concentrated in coated pits and vesicles in the basal cytoplasm. After 5 min IgA was identified in larger uncoated vesicles to which coated vesicles were observed fused (9). The larger, IgA-containing vesicles selectively migrated toward the apical end of the cells. At 15–30 min IgA-containing tubular vesicles were identified fused to the apical plasmalemma and discharging IgA from the cell. The IgA was not identified associated with either Golgi bodies or lysosomes (70). This suggested that the IgA-containing tubular vesicles did not originate from the transreticular system of the Golgi apparatus (66), which is also involved in exocytosis of cell products. The association of tubular vesicles with coated vesicles, intracellular transport, and exocytosis has been documented in other systems (13, 43).

Although many details of the mechanism need to be examined, the pathway of luteoviruses through the hindgut cell seems relatively simple. Regulation of this pathway depends on the initial recognition and binding of the virus by a membrane-bound cell receptor. At the present time we have only the ultrastructural evidence, which suggests that receptors capable of binding luteoviruses and inducing endocytosis may exist on the aphid hindgut. This important point has yet to be substantiated by biochemical or biological transmission data.

Observations of Luteovirus Transmission

In order for luteovirus transmission to occur, the virus must penetrate into, be transported through, and be released from the accessory salivary gland into the salivary duct. To better describe the cellular mechanisms involved in this process, I will present recent ultrastructural observations of accessory salivary glands of *R. padi* reared 14 days on either healthy oats or on oats infected with the RPV-NY isolate of BYDV. All methods were as previously described (21, 23).

Accessory Salivary Gland Structure

The salivary glands of *R. padi* were similar in structure and location to those described for *Sitobion avenae* (23) and *Myzus persicae* (21, 56). The salivary system consisted of two sets of glands, each with a small accessory salivary gland and a larger principal salivary gland located immediately posterior to the optic lobe nerve ganglion (Fig. 4.3 and 4.4). Luteoviruses have been observed associated only with the simple four-celled accessory salivary glands. The larger, more complex principal salivary glands are not known to be directly involved in regulating luteovirus transmission. The size of the accessory salivary gland varies with the size of the aphid; however, it is often in the 50 × 100 μm size range. The accessory gland drains its secretory products into the salivary duct, which is joined by a short duct coming from the principal gland to form a common salivary duct (Fig. 4.15) leading to the salivary syringe. The syringe pumps salivary duct contents into the salivary canal of the aphid stylets. The four cells of the accessory gland are ultrastructurally similar (Fig. 4.16). Surrounding each gland is an extracellular basal lamina, which acts as a filtering barrier to any compounds diffusing from the hemocoel toward the accessory gland secretory cells. Lying below the basal lamina is the basal plasmalemma. This membrane border is highly invaginated, resulting in a large cell surface area at the hemocoel–accessory gland interface. Within the cytoplasm and closely associated with the membrane invaginations are numerous mitochondria. Just below the area of invaginations and mitochondria are accumulated a variety of vesicles, including lysosomes, multivesicular bodies, and secretory vesicles. Rough endoplasmic reticulum is common throughout the cytoplasm, and coated pits and vesicles can be observed associated with the apical and basal plasmalemma. Typical Golgi bodies consisting of *cis* and *trans* faces of stacks of cisternae (66) were not observed in the accessory glands with the fixation procedures employed (21). The extreme sensitivity of the accessory gland to the fixation protocol was demonstrated by several experiments in which principal gland, muscle, and nerve tissue cytoplasmic organelle ultrastructure were well preserved, while accessory gland cells of the same aphids showed poor preservation that resulted in swollen mitochondria and membrane disruptions. The most characteristic feature of the accessory gland cytoplasm is the microvilli-lined canal which ramifies throughout each cell. The membrane forming the canal is actually the highly invaginated apical plasmalemma of each cell. The lumen of the canal is in direct continuity with the lumen of the chitin-lined salivary duct, as described for *M. persicae* (21). To be transmitted, luteoviruses suspended in hemolymph must diffuse through the

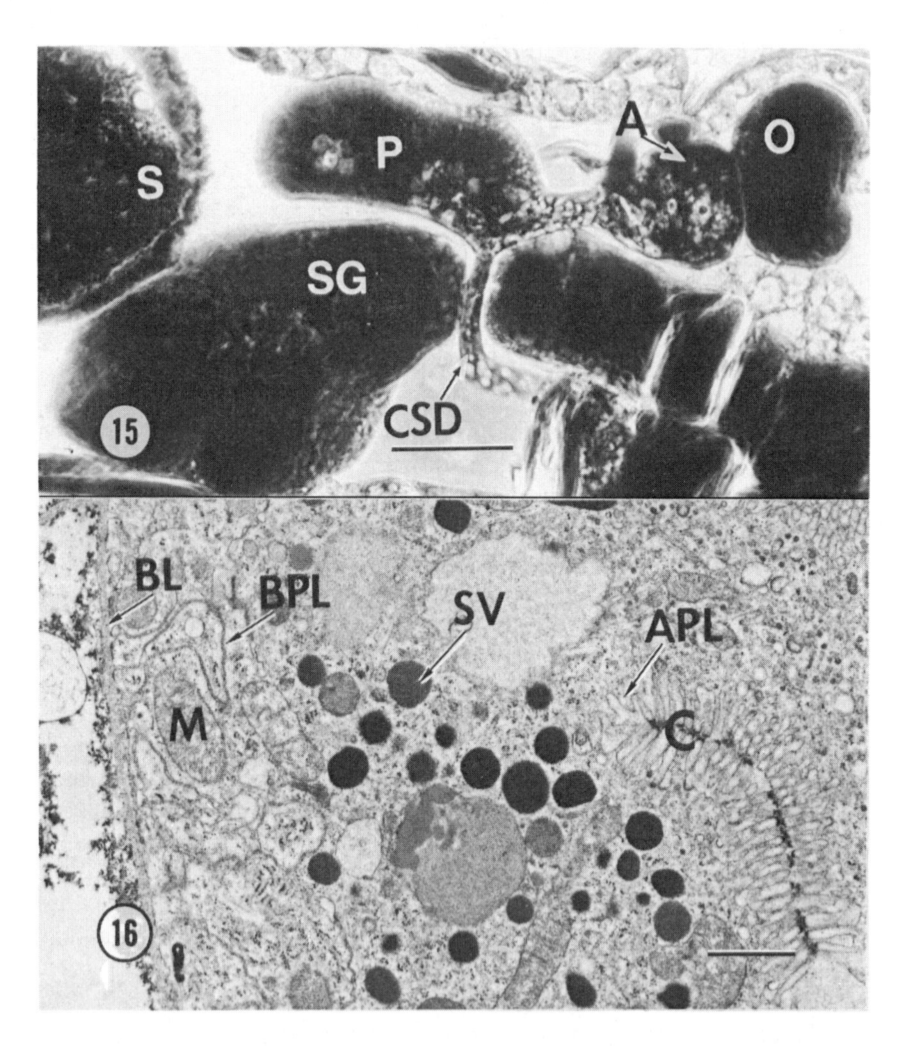

FIGURES 4.15–4.16. Accessory salivary gland structure of *Rhopalosiphum padi*. FIGURE 4.15. Light micrograph of a tangential section showing the association of the accessory salivary gland (A) and principal salivary gland (P) and their connection to the common salivary duct (CSD), optic lobe (O), subesophegeal nerve ganglion (SG), and stomach region of midgut (S). Bar = 50 μm. FIGURE 4.16. Electron micrograph of an accessory gland cell showing the extracellular basal lamina (BL), invaginated basal plasmalemma (BPL), mitochondria (M), secretory vesicles (SV), and apical plasmalemma (APL), which forms the microvilli-lined canals (C). Bar = 1μm.

basal lamina, be actively transported through the basal plasmalemma and across the cytoplasm to the apical end of the cell, and be released into the lumen of the microvilli-lined canal. From this location there are no further barriers to transmission.

Barley Yellow Dwarf–Accessory Salivary Gland Interactions

In *R. padi* reared on RPV-infected oats, virus particles were consistently observed associated with the accessory salivary gland in 17 of 20 aphids examined by transmission electron microscopy. Virions were most easily seen when embedded in the basal lamina or within the exocytoplasmic space of the basal plasmalemma invaginations (Fig. 4.17). Virions sometimes appeared like strings of beads linearly arrayed in membrane invaginations (Fig. 4.18). In serial sections these arrays most frequently occurred near tips of the invaginations. In the central cytoplasm, associated with the region of secretory vesicles, virions were observed in tubular vesicles similar in appearance to those described in the hindgut (Fig. 4.19). Virions were not observed to be concentrated in lysosomes, receptosomelike smooth spherical vesicles, or secretory vesicles. Within the cytoplasm the most numerous accumulations of virions occurred in coated vesicles in the apical cytoplasm adjacent to canals (Fig. 4.20) and in coated pits fused to the canal membrane. These observations confirm and extend the range of ultrastructural associations of luteoviruses with the accessory gland cytoplasm previously described for other luteovirus–aphid systems (21, 23). The associations of RPV with cytoplasmic tubular vesicles and linearly arrayed in basal plasmalemma invaginations were much more evident in the RPV–*R. padi* system than in others previously observed. These virus–membrane associations suggest possible modes of luteovirus transport through the salivary cell.

Immunocytochemical Identification of RPV

Immunocytochemistry was used to verify the identity of the virus particles observed. This was necessary in order to avoid confusion with other possible luteoviruses which could contaminate the plant host tissue and to demonstrate that the observed particles were not an aphid virus. In one experiment heads of viruliferous *R. padi* fed on RPV-infected oats were dissected in 0.05 M phosphate buffer (pH 7.0) to expose the salivary glands, which were loosely attached to the subesophageal nerve ganglion. These tissues were incubated overnight at 4°C in the IgG fraction of rabbit polyclonal antibody made against the RPV-NY or MAV-NY isolates of BYDV (59) diluted to 0.2 mg/ml. After rinsing in buffer, the tissues were incubated in 0.1 mg/ml ferritin-conjugated, goat anti-rabbit IgG for 3 hr at 4°C. Tissues were then rinsed in buffer four times over 1 hr and fixed for TEM as previously described (21). Virus particles embedded in the

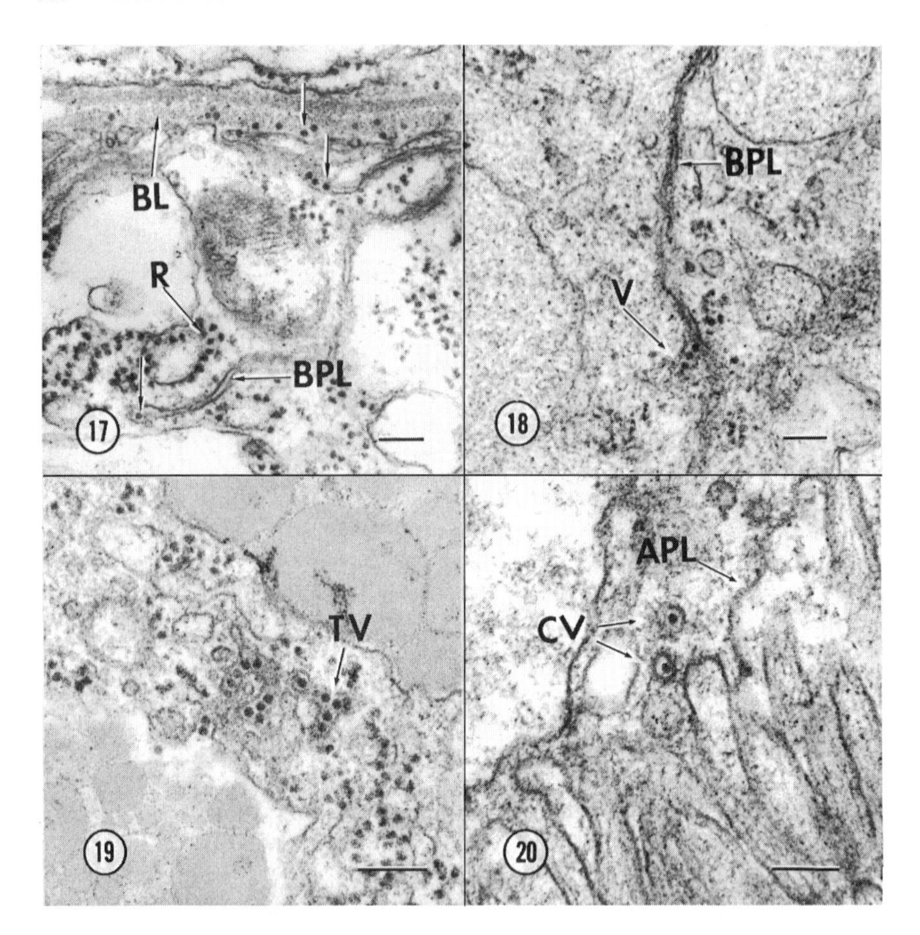

FIGURES 4.17–4.20. Electron micrographs illustrating the ultrastructural association of virions of the RPV isolate of BYDV in accessory salivary gland cytoplasm of *Rhopalosiphum padi*. FIGURE 4.17. Virions (unlabeled arrows) embedded in the basal lamina (BL) and within the lumen of basal plasmalemma (BPL) invaginations; cytoplasmic ribosomes (R). FIGURE 4.18. Virions (V) in a tubular array associated with the plasmalemma (BPL) invaginated deep into the cytoplasm. FIGURE 4.19. Virions in membrane-bound tubular vesicles (TV) adjacent to secretory vesicles. FIGURE 4.20. Virions in coated vesicles (CV) adjacent to the apical plasmalemma forming the microvilli-lined canals. Bars = 200 nm.

fibrous basal lamina (Fig. 4.21) and in the lumen of the plasmalemma invaginations of the accessory gland were positively labeled in all three aphids examined that were treated with RPV antiserum. Labeled particles were not found associated with the principal salivary gland, muscle, optic lobe of the brain, or subesophageal nerve ganglion. No serological labeling of particles occurred in three aphids treated with heterologous MAV antiserum.

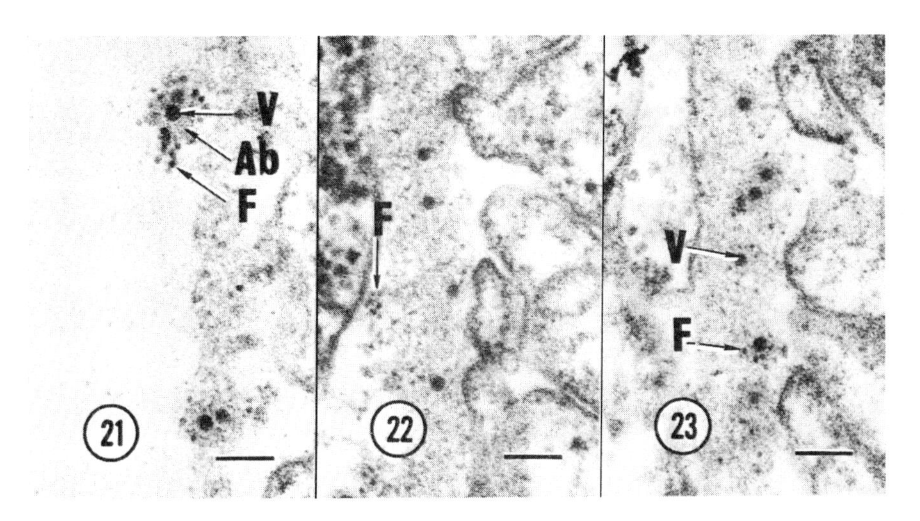

FIGURE 4.21–4.23. Virions of RPV associated with accessory salivary gland basal lamina of *Rhopalosiphum padi*. FIGURE 4.21. Labeled virions in basal lamina of an accessory gland dissected from a viruliferous *R. padi* and incubated in anti-RPV antiserum followed by ferritin-conjugated goat anti-rabbit antiserum. Note coating of virions (V) by antibodies (Ab) and ferritin (F). FIGURE 4.22. Unlabeled RPV virions in basal lamina from *R. padi* previously microinjected with antiserum to the MAV isolate of BYDV followed by ferrin-conjugated goat anti-rabbit antiserum. FIGURE 4.23. Ferritin (F)-labeled virions (V) from an aphid treated as above, but injected with anti-RPV antiserum. Bar = 100 nm.

To avoid the tedious procedure of dissecting aphid tissues and to avoid the loss of ultrastructural detail resulting from tissue incubations in buffer, another method for labeling virions *in vivo* was developed. In this procedure, *R. padi* fed 10 days on RPV-infected oats were microinjected with 0.01 μl of rabbit IgG (0.2 mg/ml) made against RPV or MAV, or against RhPV, an aphid virus (12). The aphids were allowed to feed overnight and then injected the next day with the ferritin-conjugated, goat anti-rabbit IgG at 0.1 mg/ml. The aphids were fixed 6 hr later for TEM. Virions embedded in the accessory gland basal lamina were not labeled in any of three aphids injected with anti-MAV (Fig. 4.22) or anti-RhPV antiserum. Ferritin particles were observed scattered randomly in the hemocoel and associated with the basal lamina in these treatments, indicating the labeled antibody did have access to the virus. In three aphids injected with anti-RPV, the virions were specifically labeled with a coating of antibody and ferritin, as seen in Figure 4.23. Injection of the homologous RPV antiserum resulted in an accumulation of labeled particles in the accessory gland basal lamina, but not in basal lamina of adjacent principal gland or nerve tissues. These results suggest an extreme specificity for the association of the luteovirus to the accessory salivary gland basal lamina. Antibody injection had no effect on the normal cytoplasmic ultrastructure. Injection of anti-RPV, however, did result in the aggregation of virions in the lumen of the microvilli-lined canals and an increased accumulation of particles in accessory gland cytoplasm within arrays of tubular vesicles. Such accumulations did not occur with the anti-MAV or RhPV antisera. These observations suggest that the homologous antiserum was transported into the canal lumen, possibly through the tubular vesicle system, where it was then able to react with RPV accumulating in tubular vesicles or being released into the canal. Ferritin molecules did not penetrate the cell membrane and ferritin labeling did not occur beyond the basal plasmalemma. The *in vivo* labeling and antiserum-induced virus accumulation in the gland greatly facilitated virus visualization and might be an excellent method for localizing difficult to identify virus of low concentrations.

Enzyme Histochemical Localization of RPV

How luteoviruses move through the basal plasmalemma to accumulate in tubular vesicles in the central or apical cytoplasm is unknown. Associations of virions with coated pits on the basal membrane have not been observed and virion-containing coated vesicles have only been observed a few times in the basal cytoplasm (21). Therefore, it is unlikely that endocytosis into coated vesicles is involved. Whether or not luteoviruses could penetrate the membrane and occur free in the cytoplasm is difficult to determine due to the similarity in size (20–25 nm) and electron density characteristics of luteoviruses and cytoplasmic ribosomes. To determine whether or not RPV could exist as free particles in the accessory gland cytoplasm, an enzyme histochemical test (32) was done to eliminate cytoplasmic ribosomes before tissues were embedded for TEM. Salivary glands were dissected as previously described from viruliferous *R. padi* fed 10 days on RPV-infected oats. The dissected glands were fixed 15 min in 1.0% glutaraldehyde in 0.01 M phosphate buffer (pH 7.0) on ice and then rinsed five times in buffer over 60 min. The glands were then incubated overnight in either 0.01 M phosphate buffer or in 0.1 mg/ml RNase (Sigma type III) in buffer at 20°C. After the RNase treatment the tissues were rinsed in buffer, fixed, and embedded for TEM (21).

The cell ultrastructure was reasonably well preserved in tissues fixed in 1.0% glutaraldehyde and then incubated in buffer. Ribosomes were evident throughout the cytoplasm and RPV virions were observed in the basal lamina, membrane invaginations (Fig. 4.24), and coated vesicles near the microvilli-lined canals. In glands of five aphids incubated in RNase (Figs. 4.25–4.27) virions were well preserved, but ribosomes had been enzymatically degraded and were not observed. This allowed easy identification of virus throughout the cell. Although cell ultrastructure was not as well preserved following the RNase treatment, virions were only observed to occur within discrete membrane-bound structures similar to tubular vesicles and coated vesicles in accessory gland cytoplasm. In ad-

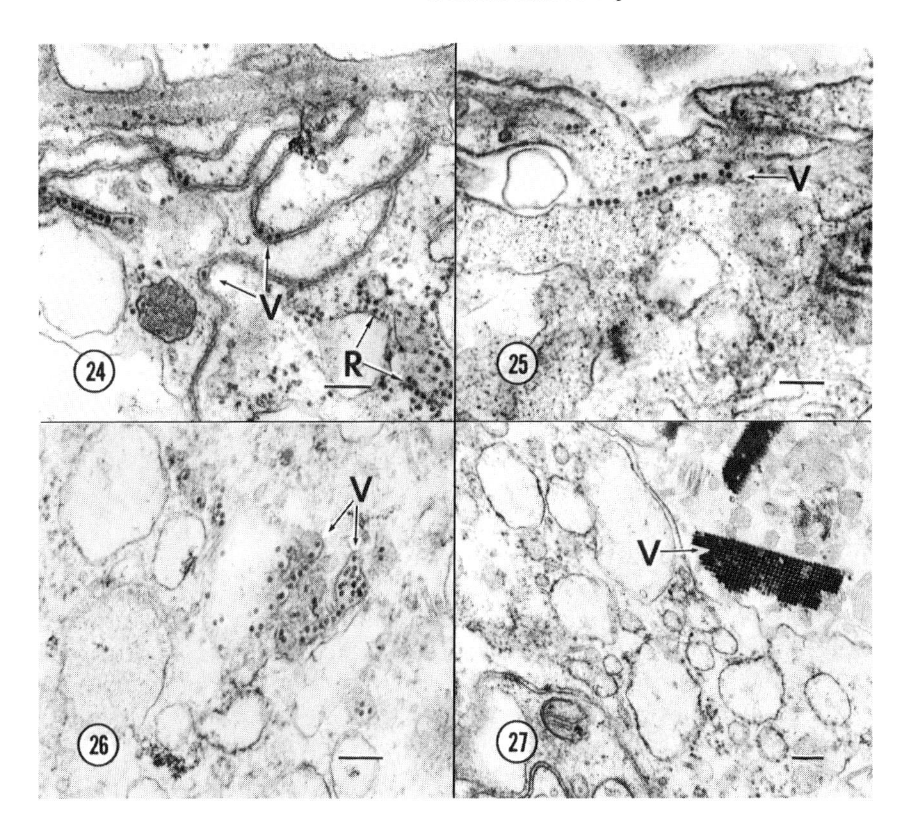

FIGURES 4.24–4.27. Virions of the RPV isolate of BYDV in accessory salivary glands of *Rhopalosiphum padi*. FIGURE 4.24. Electron micrograph of an accessory gland dissected from a viruliferous *R. Padi* and incubated in phosphate buffer. Note presence of virions (V) associated with membrane invaginations and cytoplasmic ribosomes (R). FIGURES 4.25–4.27. Similar to Figure 4.24, but incubated in RNase. Note absence of ribosomes. FIGURE 4.25. Virions in basal lamina and lumen of membrane invagination. FIGURE 4.26. Virions in membrane-bound structures in the cytoplasm. FIGURE 4.27. Virions in crystalline arrays located in a lysosome. Bars = 200 nm.

dition, large crystalline arrays of particles were observed in lysosomes, indicating that large amounts of luteovirus could accumulate in the salivary gland. Similar arrays have never been observed in untreated tissues free of artifacts. These results suggest, however, that RPV does not occur free in the cytoplasm and that mechanisms of ingress and intracellular transport involve only virions packaged within cellular membranes.

A Luteovirus Transmission Model

A hypothetical model describing luteovirus movement through the aphid accessory salivary gland was previously described (21). The results reported here for RPV in *R. padi* strengthen various aspects of the model and add new insights into the membrane–virus interactions involved. At the present time the ultrastructural evidence suggests that luteoviruses suspended in the hemolymph diffuse through the fibrous accessory gland basal lamina and come into contact with the basal plasmalemma. Virus-specific receptors may then bind the virus to the membrane. Exactly how the virus is internalized is not clear. The lack of virion-containing coated vesicles in the basal cytoplasm suggests that endocytosis by coated pits is probably not involved. The consistent visualization of virions singly and in linear arrays deep within the membrane invaginations (Fig. 4.18) suggests that the virus might be endocytosed by accumulations of virions within invaginations which bud off of the plasmalemma to form smooth-membraned tubular vesicles. Such vesicles could fuse together, forming tubular vesicles (Fig. 4.19) analogous to receptosomes. Virus-containing tubular vesicles located in the apical cytoplasm near the canal are frequently associated with coated vesicles, and coated pits containing virions have been observed fused to these tubular vesicles (21). These observations suggest that virions in the tubular vesicle are packaged in a specific manner in coated vesicles for transport to the apical plasmalemma and exocytosed into the canal lumen. The role of coated vesicles in exocytosis of macromolecules is well established (53, 76). The lack of accumulation of luteovirus particles in lysosomes and secretory vesicles under normal circumstances indicates that virion transport in the coated vesicle is very site-specific. The coated vesicle migrates to the apical canal membrane (Fig. 4.20), fuses to it, and becomes a coated pit with its lumen contents released into the canal.

The proposed pathway to describe luteovirus transport through the accessory gland is ultrastructurally similar to the cellular mechanism for IgG transepithelial transport across certain mammalian intestinal cells (1). IgG in the intestinal lumen binds to specific receptors on the intestinal cell membrane. The IgG accumulates in membrane invaginations, which eventually bud off of the plasmalemma to form smooth-membraned endocytotic tubules. Coated vesicles then form from these tubules and transport the IgG to the opposite side of the cell for exocytosis into the circulatory system.

Discussion

A model for luteovirus transmission requiring virus recognition and binding to virus-specific receptors that regulate virus uptake into the accessory gland is compatible with several lines of experimental evidence.

Early work by Rochow and Pang (62) indicated that some aphids while feeding on BYDV-infected plants could acquire into their hemocoel isolates of BYDV, which they could not transmit. In addition, vector-specific transmission of BYDV isolates was maintained even when purified virus preparations were injected directly into the hemocoel (57). These results indicated that at least for some luteovirus–aphid combinations the site preventing transmission of nonvectored isolates was not associated with the alimentary canal. Since regulation occurred after virus entered the hemocoel, salivary glands were suggested as the next logical point of control (63). Microinjection bioassays of aphid tissue extracts were used by Paliwal and Sinha (50) to test for BYDV in various organs of *S. avenae* fed on infected oats. Virus was detected in aphid gut, hemolymph, and salivary glands, but not in brain tissue. No specific association of virus with the accessory gland was detected in this study, however, since both the principal and accessory salivary glands were dissected and tested together.

The accessory salivary gland as a specific site for selective regulation of circulative virus transmission by aphids was first suggested by ultrastructural studies of pea enation mosaic virus (PEMV) in *Acythrosiphon pisum,* the pea aphid (28, 29). Harris and co-workers observed virions of an aphid-transmissible isolate (T) of PEMV concentrated in the basal lamina surrounding the accessory gland and associated with the underlying plasmalemma when aphids were allowed to feed on PEMV-infected plants or when aphids were microinjected with partially purified virus preparations. Virions of an aphid-nontransmissible PEMV isolate (NT), selected by repeated mechanical inoculations, could not be detected at the accessory gland following injection of purified NT virions. This work pointed to the accessory salivary gland as a probable site involved in circulative virus transmission, and suggested that the accessory gland basal lamina could be involved in selecting which virions reached the plasmalemma (28). The importance of the accessory gland to circulative transmission was substantiated in subsequent ultrastructural studies of BYDV in the vector *S. avenae* (23). The MAV isolate of BYDV, which is specifically transmitted by *S. avenae,* was visualized embedded in the accessory gland basal lamina, in cytoplasmic coated vesicles, and in the lumen of canals draining the gland. Virions of the RPV isolate transmitted specifically by *R. padi,* but not by *S. avenae,* were also observed in *S. avenae* embedded in the basal lamina, but not in sites beyond the basal plasmalemma. The results verified that the accessory gland could be a general route for transmission of different groups of circulative viruses; and, more importantly, indicated that BYDV vector specificity might be regulated at the accessory gland plasmalemma. Similar results were observed in later studies of potato leaf roll virus and beet western yellows virus transmitted by *M. persicae* (21).

The above studies indicated that the accessory gland was a site where regulation of luteovirus vector specificity could occur. The molecular mechanism for cellular regulation of transmission was suggested by another line of evidence involving genomic masking or transcapsidation by luteovirus isolates replicating together in mixed infections (58). For example,

R. padi acquires and transmits RPV, but not MAV, from plants infected with either isolate of BYDV. When *R. padi* feed on plants doubly infected with RPV and MAV, however, the aphids occasionally transmit MAV along with RPV. Serological tests (61) indicated that the MAV genome was encapsidated in the RPV coat protein during virion assembly. These continuing studies (35) have led to the current hypothesis that vector-specific transmission of luteoviruses is based on an interaction between the virus capsid protein and elements of the accessory gland cell membrane. This hypothesis is also supported by studies of PEMV (2), which described significant differences in coat protein structure between the T and NT isolates observed by Harris *et al.* (28, 29).

The idea that receptor-mediated endocytosis may regulate luteovirus transmission was supported by yet another line of evidence resulting from a fortuitous observation. When individual *S. avenae* were allowed to feed first on MAV-infected oats and then on PAV-infected oats, the aphids were less likely to transmit PAV than were aphids fed first on healthy oats (25). The inhibition of PAV transmission by prior acquisition of MAV was shown to be specific in *S. avenae* between these two isolates. The MAV and PAV isolates are serologically closely related (59), indicating some similarity in their capsid protein structure. Similar interference in transmission did not occur between PAV and RPV or RMV, which are not serologically similar, indicating significant differences in capsid protein structure. The MAV–PAV transmission interference phenomenon was also observed when purified preparations of MAV and PAV were injected simultaneously into *S. avenae,* which were then allowed to feed on plants. The level of interference depended on the amount of MAV relative to PAV in the mixture, since PAV transmission decreased as the MAV concentration increased. When purified MAV was UV-irradiated for 15–30 sec before being mixed with PAV for injection, MAV no longer interferred with PAV transmission (24). Results of ELISA tests, ultrastructural observations, and bioassays suggested that the MAV coat protein structure was altered by irradiation and MAV was either not transmitted or was not infectious. In either case, MAV was no longer able to interfere with PAV transmission. These results and observations are compatible with the hypothesis that MAV and PAV are recognized by and attach to the same population of membrane receptors on the accessory gland plasmalemma. If this is true, then saturation of such receptors with MAV would reduce the ability of PAV to be recognized and transported into the cell, resulting in the observed reduction of PAV transmission.

One argument against the above hypothesis for MAV–PAV interference as evidence of receptor regulation of transmission results from the ability of the serologically related MAV and PAV isolates to induce cross-protection against each other in plants (36). In order for cross-protection to occur, however, the protecting isolate must be inoculated 4 days prior to the challenge inoculation. Our observation of MAV–PAV interference following simultaneous injection of both isolates into aphids would seem to preclude the possibility of MAV consistently establishing itself and cross-protecting against PAV replication at a consistent level in all experiments over a 6-year period. To test whether reduced PAV recovery from plants fed on by *S. avenae* previously exposed to MAV and PAV could result from MAV becoming established first and inducing cross-protection, oats were inoculated simultaneously with MAV and PAV or with PAV alone (25). The presence of MAV had no effect on the recovery of PAV from doubly infected plants when compared to recovery from plants inoculated only with PAV. This further supported the idea that the MAV–PAV interference occurred in the aphid salivary gland.

More recent work has shown that PAV does not interfere with MAV transmission in a reciprocal manner when tested under similar experimental parameters (64). This suggests that the receptors, if they exist, have a greater binding affinity for MAV than for PAV. This is compatable with the fact that *S. avenae* transmits MAV much more efficiently than PAV in single aphid tests. No interference of PAV transmission by RPV occurred in *R. padi* even though *R. padi* transmits both isolates very efficiently (25). These two isolates may utilize distinct populations of receptors on the *R. padi* accessory gland membrane, since they are serologically not closely related (59). The occurrence of distinct populations of receptors on cell membranes recognizing different virus isolates has been described for other virus–host systems (10, 17).

Future Perspectives

Much indirect evidence has accumulated to suggest that aphid membrane receptors regulate luteovirus acquisition and transmission. Most of this evidence is based on various types of bioassays, aphid transmission tests, and ultrastructural studies. Methods must now be developed to make direct tests for the existence of such receptors. To definitively answer the question, "Do receptors exist?," the proposed receptors must be identified on the membrane, purified, and characterized. To better understand the virus–receptor interaction leading to virus recognition and transport through the aphid will also require a better understanding of luteovirus composition and structure. At the present time the luteoviruses are relatively poorly characterized. Of primary concern will be the protein structure of the capsid which presents the recognition site for receptor attachment. Developing methodologies should be able to overcome the difficulties associated with luteovirus–aphid systems, such as the small size of the aphid tissue and the limited concentration of virus involved. Antiidiotype antibodies made against luteovirus-specific antibody (34) could be used with immunoelectron microscopy to identify receptor proteins on the cell membrane and for use with protein A chromatography for receptor purification (68). Monoclonal antibody studies of viral epitope differences among vector-specific isolates could give information concerning the viral recognition site. Capsid protein characterization and amino acid sequencing would provide information concerning protein structure involved in the spatial alignment of specific regions, allowing the virus–receptor interaction to occur and would indicate the extent of virus modification necessary for luteoviruses to acquire new vector species. Luteovirus-receptor studies could also be greatly advanced by the development of continuous aphid cell cultures. To date methods have not been developed to allow the continuous maintenance of secondary aphid cell lines (3). If aphid cell lines could be developed and selected for expression of luteovirus receptors, then this would allow virus interference and binding kinetics studies (10) not currently possible.

Although much has been accomplished toward describing circulative virus transmission, much remains to be done to understand it. Developing technologies should make this goal possible in the foreseeable future. Increasing our knowledge of the cellular and molecular mechanisms regulating luteovirus transmission will give us a better understanding of insect transmission of viruses of agricultural and medical importance, and should lead toward the development of innovative control strategies in the future.

Acknowledgments. I would like to thank Dr. William F. Rochow for his continuous encouragement and support of my research program over the past 10 years. His persistence and devotion to science has provided a basis for our understanding of luteovirus biology. This research was funded by grants from the Rockefeller Foundation to the author.

References

1. Abrahamson, D.R., and Rodewald, R., 1981, Evidence for the sorting of endocytotic vesicle contents during the receptor-mediated transport of IgG across the newborn rat intestine, *J. Cell Biol.* **91:**270–280.
2. Adam, G., Sander, E., and Shepherd, R.J., 1979, Structural differences between pea enation mosaic virus strains affecting transmissibility by *Acyrthosiphon pisum* (Harris), *Virology* **92:**1–14.
3. Adam, G., Sander, E., and Lee, P.E., 1980, Inoculation of aphid primary cultures with a iodinated plant virus, in: E. Kurstak, K. Maramorosch, and A. Dubendorfer (eds.), *Invertebrate Systems in Vitro,* Elsevier/North-Holland Press, Amsterdam, pp. 485–492.
4. Armour, S.L., Melcher, U., Pirone, T.P., Little, D.J., and Essenberg, R.D., 1983, Helper component for aphid transmission encoded by region II of cauliflower mosaic virus DNA, *Virology* **129:**25–30.
5. Bowers, B., 1964, Coated vesicles in the pericardial cells of the aphid (*Myzus persicae* Sulz.), *Protoplasma* **59:**351–362.

6. Brown, M.S., and Goldstein, J.L., 1986, A receptor-mediated pathway for cholesterol homeostasis, *Science* **232:**34–47.
7. Clark, R.G., and Bath, J.E., 1973, Transmission of pea enation mosaic virus by the pea aphid, *Acyrthosiphon pisum,* following virus acquisition by injection, *Ann. Entomol. Soc. Am.* **66:**603–607.
8. Cockbain, A.J., and Costa, C.L., 1973, Comparative transmission of bean leaf roll and pea enation mosaic viruses by aphids, *Ann. Appl. Biol.* **73:**167–176.
9. Courtoy, P.J., Limet, J.N., Quintart, J., Schneider, Y.J., Vaerman, J.P., and Baudhuin, P., 1983, Transport of IgA into rat bile: Ultrastructural demonstration, *Ann. N.Y. Acad. Sci.* **409:**799–802.
10. Crowell, R.L., 1976, Comparative generic characteristics of picornavirus–receptor interactions, in: R.F. Beers and E.G. Bassett (eds.), *Cell Membrane Receptors for Viruses, Antigens, Antibodies, Polypeptide Hormones, and Small Molecules,* Raven Press, New York, pp. 179–202.
11. Dales, S., 1978, Penetration of animal viruses into cells, in: S.S. Silverstein (ed.), *Transport of Macromolecules in Cellular Systems,* Dahlem Konferenzen, Berlin, pp. 47–68.
12. D'Arcy, C.J., Burnett, P.A., Hewings, A.D., and Goodman, R.M., 1981, Purification and characterization of a virus from the aphid *Rhopalosiphum padi, Virology* **112:**346–349.
13. Debruyn, P.P., Michelson, S., and Becker, R.P., 1975, Endocytosis, transfer tubules, and lysosomal activity in myeloid sinusoidal endothelium, *J. Ultrastruct. Res.* **53:**133–151.
14. Duffus, J.E., 1960, Radish yellows, a disease of radish, sugar beet, and other crops, *Phytopathology* **50:**389–394.
15. Duffus, J.E., 1979, Legume yellows virus, a new persistent aphid-transmitted virus of legumes in California, *Phytopathology* **69:**217–221.
16. Elnagar, S., and Murant, A.F., 1978, Relations of carrot red leaf and carrot mottle viruses with their aphid vector, *Cavariella aegopodii, Ann. Appl. Biol.* **89:**237–244.
17. Epstein, R.L., Powers, M.L., Rogart, R.B., and Weiner, H.L., 1984, Binding of I^{125}-labeled reovirus to cell receptors, *Virology* **133:**46–55.
18. Eskandari, F., Sylvester, E.S., and Richardson, J., 1979, Evidence for lack of propagation of potato leafroll virus in its aphid vector, *Myzus persicae, Phytopathology* **69:**45–47.
19. Fitzgerald, D.P., Padmanabhan, R., Pastan, I., and Willingham, M.C., 1983, Adenovirus-induced release of epidermal growth factor and *Pseudomonas* toxin into the cytosol of KB cells during receptor-mediated endocytosis, *Cell* **32:**607–617.
20. Gera, A., Loebenstein, G., and Raccah, B., 1979, Protein coats of two strains of cucumber mosaic virus affect transmission by *Aphis gossypii, Phytopathology* **69:**396–399.
21. Gildow, F.E., 1982, Coated-vesicle transport of luteoviruses through salivary glands of *Myzus persicae, Phytopathology* **72:**1289–1296.
22. Gildow, F.E., 1985, Transcellular transport of barley yellow dwarf virus into the hemocoel of the aphid vector, *Rhopalosiphum padi, Phytopathology* **75:**292–297.

23. Gildow, F.E., and Rochow, W.F., 1980, Role of accessory salivary glands in aphid transmission of barley yellow dwarf virus, *Virology* **104:**97–108.
24. Gildow, F.E., and Rochow, W.F., 1980, Importance of capsid integrity for interference between two isolates of barley yellow dwarf virus in an aphid, *Phytopathology* **70:**1013–1015.
25. Gildow, F.E., and Rochow, W.F., 1980, Transmission interference between two isolates of barley yellow dwarf virus in *Macrosiphum avenae, Phytopathology* **70:**122–126.
26. Hanover, J.A., and Dickson, R.B., 1985, Transferrin: Receptor-mediated endocytosis and iron delivery, in: I. Pastan and M.C. Willingham (eds.), *Endocytosis,* Plenum Press, pp. 131–161.
27. Harris, K.F., 1977, An ingestion–egestion hypothesis of noncirculative virus transmission, in: K.F. Harris and K. Maramorosch (eds.), *Aphids as Virus Vectors,* Academic Press, New York, pp. 162–220.
28. Harris, K., 1979, Leafhoppers and aphids as biological vectors: Vector–virus relationships, in: K. Maramorosch and K.F. Harris (eds.), *Leafhopper Vectors and Plant Disease Agents,* Academic Press, New York, pp. 217–308.
29. Harris, K.F., Bath, J.E., Thottappilly, G., and Hooper, G.R., 1975, Fate of pea enation mosaic virus in PEMV-infected pea aphids, *Virology* **65:**148–162.
30. Harrison, B.D., 1958, Studies on the behavior of potato leaf roll and other viruses in the body of their aphid vector *Myzus persicae* (Sulz.), *Virology* **6:**265–277.
31. Harrison, B.D., and Murant, A.F., 1984, Involvement of virus-coded proteins in transmission of plant viruses by vectors, in: M.A. Mayo and K.A. Harrap (eds.), *Vectors in Virus Biology,* Academic Press, London, pp. 1–36.
32. Hatta, T., and Francki, R.I.B., 1979, Enzyme cytochemical methods for identification of cucumber mosaic virus particles in infected cells, *Virology* **93:**265–268.
33. Helenius, A., Kartenbeck, J., Simons, K., and Fries, E., 1980, On the entry of Semliki Forest virus into BHK-21 cells, *J. Cell Biol.* **84:**404–420.
34. Hu, J.S., and Rochow, W.F., 1986, Production of an anti-idiotype antibody to an anti-barley yellow dwarf virus monoclonal antibody, *Phytopathology* **76:**1062 (abstract).
35. Hu, J., Rochow, W.F., and Dietert, R.R., 1985, Production and use of antibodies from hen eggs for the SGV isolate of barley yellow dwarf virus, *Phytopathology* **75:**914–919.
36. Jedlinski, H., and Brown, C.M., 1965, Cross protection and mutual exclusion by three strains of barley yellow dwarf virus in *Avena sativa* L., *Virology* **26:**613–621.
37. Jensen, S.G., 1969, Occurrence of virus particles in the phloem tissue of BYDV-infected barley, *Virology* **38:**83–91.
38. Jensen, S.G., 1972, Metabolism and carbohydrate composition in barley yellow dwarf virus-infected wheat, *Phytopathology* **62:**587–592.
39. Johnson, R.A., and Rochow, W.F., 1972, An isolate of barley yellow dwarf virus transmitted specifically by *Schizaphis graminum, Phytopathology* **62:**921–925.
40. Kefalides, N.A., 1978, Current status of chemistry and structure of basement membranes, in: N.A. Kefalides (ed.), *Biology and Chemistry of Basement Members,* Academic Press, New York, pp. 215–228.

41. Kraehenbuhl, J.P., and Kuhn, L., 1978, Transport of immunoglobulins across epithelia, in: S.C. Silverstein (ed.), *Transport of Macromolecules in Cellular Systems,* Dahlem Konferenzen, Berlin, pp. 213–228.
42. Leonard, S.H., and Holbrook, F.R., 1978, Minimum acquisition and transmission times for potato leafroll virus by the green peach aphid, *Ann. Entomol. Soc. Am.* **71**:493–495.
43. Levine, J.S., Nakene, P.K., and Allen, R.H., 1981, Human intrinsic factor secretion: Immunocytochemical demonstration of membrane-associated vesicular transport in parietal cells, *J. Cell Biol.* **90**:664–655.
44. Miles, P., 1972, The saliva of hemiptera, *Adv. Insect Physiol.* **9**:183–256.
45. Mossop, D.W., and Francki, R.I.B., 1977, Association of RNA 3 with aphid transmission of cucumber mosaic virus, *Virology* **81**:177–181.
46. Murant, A.F., Robert, I.M., and Elnagar, S., 1976, Association of virus-like particles with the foregut of the aphid *Cavariella aegopodii* transmitting the semi-persistent viruses anthriscus yellows and parsnip yellow fleck, *J. Gen. Virol.* **31**:47–57.
47. Nagura, H., Nakane, P.K., and Brown, W.R., 1979, Translocation of dimeric IgA through neoplastic colon cells *in vitro, J. Immunol.* **123**:2359–2368.
48. Nault, L.R., Gyrisco, G.G., and Rochow, W.F., 1964, Biological relationship between pea enation mosaic virus and its vector, the pea aphid, *Phytopathology* **54**:1269–1272.
49. O'Loughlin, G.T., and Chambers, T.C., 1972, Extracellular microtubules in the aphid gut, *J. Cell Biol.* **53**:575–578.
50. Paliwal, Y.C., and Sinha, R.C., 1970, On the mechanism of persistence and distribution of barley yellow dwarf virus in an aphid vector, *Virology* **42**:668–680.
51. Pastan, I., and Willingham, M.C., 1981, Journey to the center of the cell: Role of the receptosome, *Science* **214**:504–509.
52. Pastan, I., and Willingham, M.C., 1985, The pathway of endocytosis, in: I. Pastan and M.C. Willingham (eds.), *Endocytosis,* Plenum Press, New York, pp. 1–44.
53. Pearse, B.M.F., and Bretscher, M.S., 1981, Membrane recycling by coated vesicles, *Annu. Rev. Biochem.* **50**:85–101.
54. Petersen, O.W., and van Deurs, B., 1983, Serial-section analysis of coated pits and vesicles involved in adsorptive pinocytosis in cultured fibroblasts, *J. Cell Biol.* **96**:277–281.
55. Pirone, T.P., and Harris, K.F., 1977, Nonpersistent transmission of plant viruses by aphids, *Annu. Rev. Phytopathol.* **15**:55–73.
56. Ponsen, M.B., 1977, Anatomy of an aphid vector: *Myzus persicae,* in: K.F. Harris and K. Maramorosch (eds.), *Aphids as Virus Vectors,* Academic Press, New York, pp. 63–82.
57. Rochow, W.F., 1969, Biological properties of four isolates of barley yellow dwarf virus, *Phytopathology* **59**:1580–1589.
58. Rochow, W.F., 1977, Dependent virus transmission from mixed infections, in: K.F. Harris and K. Maramorosch (eds.), *Aphids as Virus Vectors,* Academic Press, New York, pp. 253–273.
59. Rochow, W.F., and Carmichael, L.E., 1979, Specificity among barley yellow dwarf viruses in enzyme immunosorbent assays, *Virology* **95**:415–420.
60. Rochow, W.F., and Duffus, J.E., 1981, Luteoviruses and yellows diseases,

in: E. Kurstak (ed.), *Handbook of Plant Virus Infections and Comparative Diagnosis,* Elsevier/North-Holland, Amsterdam, pp. 147–170.
61. Rochow, W.F., and Muller, I., 1975, Use of aphids injected with virus-specific antiserum for study of plant viruses that circulate in vectors, *Virology* **63:**282–286.
62. Rochow, W.F., and Pang, E., 1961, Aphids can acquire strains of barley yellow dwarf virus they do not transmit, *Virology* **15:**382–384.
63. Rochow, W.F., Foxe, M.J., and Muller, I., 1975, A mechanism of vector-specificity for circulative aphid-transmitted plant viruses, *Ann. N.Y. Acad. Sci.* **266:**293–301.
64. Rochow, W.F., Muller, I., and Gildow, F.E., 1983, Interference between two luteoviruses in an aphid: Lack of reciprocal competition, *Phytopathology* **73:**919–922.
65. Roth, T.F., and Porter, K.R., 1964, Yolk protein uptake in the oocyte of the mosquito *Aedes aegypti* L., *J. Cell Biol.* **20:**313–332.
66. Rothman, J.E., 1981, The Golgi apparatus: Two organelles in tandem, *Science* **213:**1212–1219.
67. Shepardson, S., Esau, K., and McCrum, R., 1980, Ultrastructure of potato leaf phloem infected with potato leafroll virus, *Virology* **105:**379–392.
68. Slack, S.A., and Rochow, W.F., 1984, Molarity and pH effects on five barley yellow dwarf virus isolates, *Phytopathology* **74:**801.
69. Sylvester, E.S., 1980, Circulative and propagative virus transmission by aphids, *Annu. Rev. Entomol.* **25:**257–286.
70. Takahashi, I., Nakane, P.K., and Brown, W.R., 1982, Ultrastructural events in the translocation of polymeric IgA by rat hepatocytes, *J. Immunol.* **128:**1181–1187.
71. Tamada, T., and Harrison, B.D., 1981, Quantitative studies on the uptake and retention of potato leafroll virus by aphids in laboratory and field conditions, *Ann. Appl. Biol.* **98:**261–276.
72. Tardieu, M., Epstein, R.L., and Weiner, H.L., 1982, Interaction of viruses with cell surface receptors, *Int. Rev. Cytol.* **80:**27–61.
73. Volkman, L., and Goldsmith, P.A., 1985, Mechanism of neutralization of budded *Autographa californica* nuclear polyhedrosis virus by a monoclonal antibody: Inhibition of entry by adsorptive endocytosis, *Virology* **143:**185–195.
74. Wall, D.A., Wilson, G., and Hubbard, A.L., 1980, The galactose-specific recognition system of mammalian liver: The route of ligand internalized in rat hepatocytes, *Cell* **21:**79–93.
75. Willingham, M.C., and Pastan, I., 1983, Formation of receptosomes from plasma membrane coated pits during endocytosis: Analysis by serial sections with improved membrane labelling and preservation techniques, *Proc. Natl. Acad. Sci. USA* **80:**5617–5621.
76. Willingham, M.C., and Pastan, I., 1984, Endocytosis and exocytosis: Current concepts of vesicle traffic in animal cells, *Int. Rev. Cytol.* **92:**51–92.

5
Ecology and Evolution of Leafhopper–Grass Host Relationships in North American Grasslands

Robert F. Whitcomb, James Kramer, Michael E. Coan, and Andrew L. Hicks

". . . systematists . . . approach closer and closer to the actual truth about species separation and about phylogeny. This will give us a firmer and firmer basis for other aspects of scientific understanding based on phylogeny. I refer here to geographic dispersal, ecological dispersal, the evolution of ecological communities, and a myriad of other phenomena that cannot be related at present on their own information content but can often be related if a family tree of their taxonomic entities is available."

Herbert H. Ross (116).

Introduction

In this chapter we summarize current knowledge concerning ecological and evolutionary relationships between leafhoppers and North American grasslands. Although the treatment synthesizes information that has accumulated over a half century, the emphasis is on recent work in our own laboratory. Despite early recognition of the basic importance of host selection in insect ecology (27, 38), efforts to synthesize this information have been fragmentary because the North American grasslands that existed before European colonization were immense in size and complexity. The cicadellids that evolved with them have a similarly immense diversity of

---

Robert F. Whitcomb, Insect Pathology Laboratory, Agricultural Research Service, U.S. Department of Agriculture, Beltsville, Maryland 20705, USA.
James Kramer, Systematic Entomology Laboratory, Agricultural Research Service, U.S. Department of Agriculture, Washington, D.C. 20036, USA.
Michael E. Coan, Hemoparasitic Disease Research Unit, Agricultural Research Service, U.S. Department of Agriculture, Pullman, Washington 99164, USA.
Andrew L. Hicks, Insect Pathology Laboratory, Agricultural Research Service, U.S. Department of Agriculture, Beltsville, Maryland 20705, USA.

host relationships and life history strategies. The physical intermixing of prairie grasses and forbs is also complex, making it difficult to collect unambiguous host records, as evidenced by previous studies on cicadellid–grass relationships (31–33, 60, 63–65, 99–102, 144, 145). Once satisfactory methods were found, the task of gathering sufficient information proved to be time-consuming. Thus, even a preliminary analysis of grassland–insect communities has required many years of study.

At the time of European colonization, grasslands of various kinds were situated in central North America from the foothills of the Rocky Mountains to the edge of the unbroken deciduous forest in Ohio, from the Canadian prairie provinces to the Texan subtropics, and in various coastal and montane regions. European settlement substantially modified the boundaries of these grasslands. For example, substantial areas of the eastern forest were cut to make way for new prairie, savanna, and ecotonal habitats that were reminiscent structurally, if not in species composition, of more western grasslands and savannas. Our studies, therefore, extend from the desert plains and prairie to the eastern seaboard, simply because there has been no other obvious boundary to delimit the study area.

In this study we have considered only the Cicadellidae. In so doing, we recognize that one can ignore other grassland insect taxa only with great risk. Studies of "guilds" of auchenorrhynchous insects on single hosts, such as *Spartina patens* (37), demonstrate that valuable information is certain to be lost from such omissions. We also found it necessary to ignore the vast assemblage of cicadellids that feed wholly, or almost so, on woody plants. Some of these insects, of course, contribute to the diversity of cicadellids in savanna or prairie–forest ecotonal habitats, and the strong pattern of host specificity observed in such tree feeders (17, 18) ensures an important contribution to regional species richness. However, we are left, even after such exclusions, with cicadellid assemblages of several hundred grass species and several hundred major species of broad-leaved forbs. Extension of our study beyond these limits would surely have prolonged, postponed, or possibly even canceled this report.

We now wish to present a preliminary synthesis of the important factors in host selection and distribution of grassland cicadellids. Our conclusions are no less complex than the subject and are given in brief form in the final section of the chapter.

Historical

The beginnings of field observations on North American grassland cicadellids are to be found in earlier observations of such workers as Ball (4), Beamer (5), Hendrickson (63–65), and Osborn (99–102). These workers were followed by DeLong, who supplemented his taxonomic observations with abundant notes on the habitat relationships of cicadellids (27, 28, 31–33). Eventually, other workers began collecting and identifying cicadellids. Each new worker brought into the area unique perspectives and backgrounds, which became integral parts of their taxonomic outlooks. For example, when P.W. Oman, in 1949, revised the Nearctic genera (97), his keen observations on cicadellid–host relationships were an integral part of his overview and were discussed explicitly. It was also during these years that ecological research increasingly emphasized communities; students under this influence (70) regarded the biology of individual taxa as less interesting than the study of interrelationships among the taxa and among trophic levels within the communities. It is therefore not surprising that the views of DeLong concerning cicadellid ecology (27, 31), although he clearly recognized the existence of some host specificity, reflected habitat and community relationships and deemphasized to some extent the importance of individual host species.

It was in this climate that H.H. Ross began his studies of grassland cicadellids. Influenced in part by his knowledge of other insect groups, but also by a comprehensive overview of evolution, Ross reemphasized the importance of finding the hosts of the insects that he studied taxonomically (113–116). In fact, Ross felt that host transfer (117) represented the most likely mechanism (115) for the extensive speciation in such leafhopper genera as *Erythroneura, Flexamia,* and *Laevicephalus*. Those of us (R.F.W. and J.K.) who were his students were therefore strongly influenced in our ecological studies to "go ye therefore and find that host." In a sense, this chapter is an epistle describing that search, and it is therefore dedicated to our mentor.

The Biomes

The North American grasslands (16) comprised a vast area of the North American continent. On a geological time scale, the origins of these grasslands can be traced to the Tertiary, with the uplifting of the Rocky Mountains. This uplift significantly altered continental precipitation patterns, creating a vast "shadow" of reduced precipitation east of the ranges that gathered precipitation on their western slopes. Thus, the prairie grasslands originated as a result of climatic factors. Once formed, they were maintained by a climate of arid winters and summers that were, depending on their location, themselves divided into seasons of varying aridity. The grasslands were therefore preglacial in origin, although during the Pleistocene glaciations, their total area contracted and substantial geographic displacement of the different grassland types occurred (115). The pre-Columbian grasslands were subject to various forms of disturbance that tended to discourage conversion of prairie into forest (68). These included (i) fire caused by lightning, (ii) fires started by pre-Columbian man, (iii) agriculture, in small areas, of pre-Columbian man, (iv) grazing by various herbivores, and (v) browsing and rubbing on woody plants by herbivores.

Grasslands (Fig. 5.1) originally extended from the foothills of the Rocky

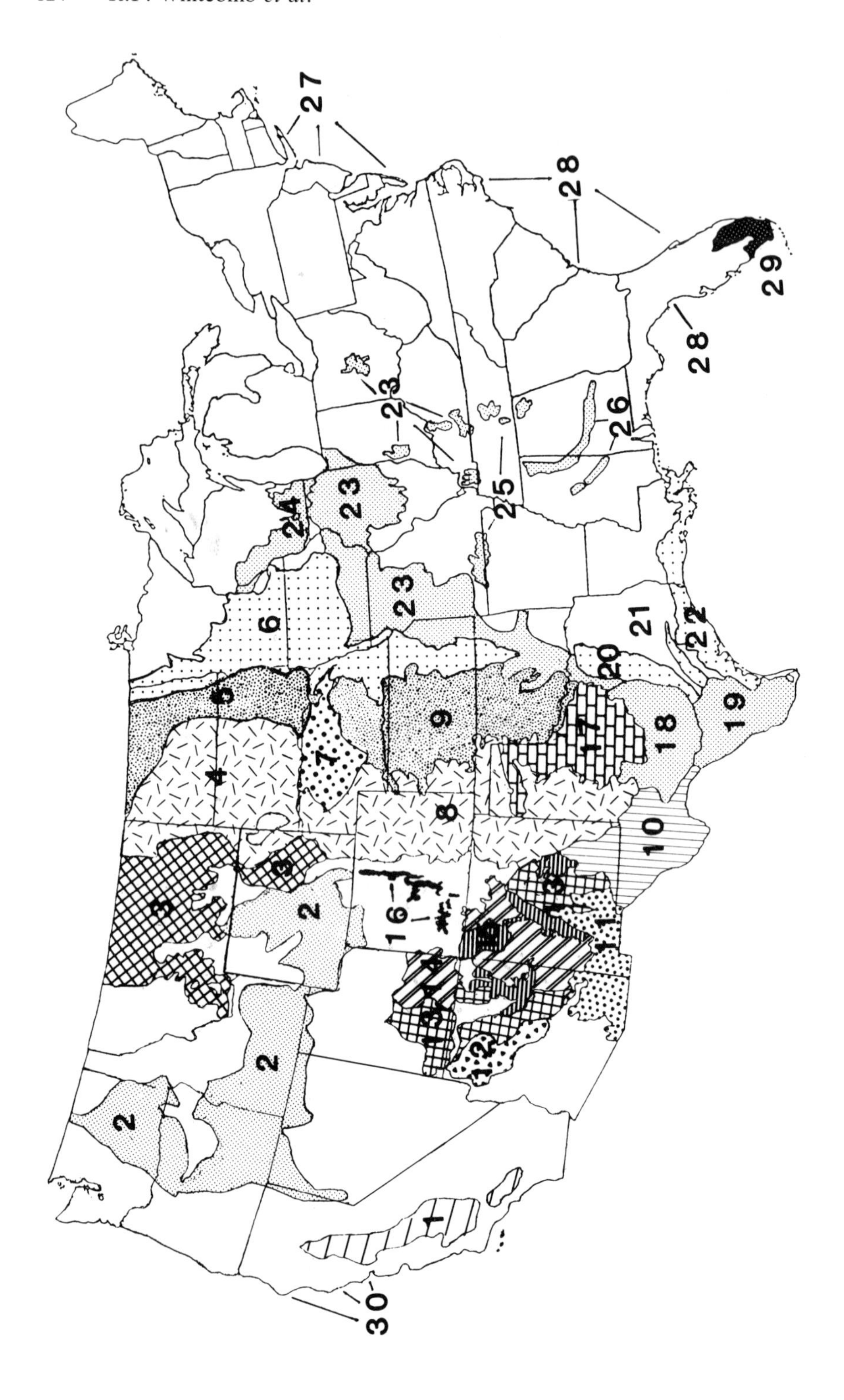
1
2
2
2
2
3
3
4
5
6
7
8
9
10
11
12
13
13
14
15
16
17
18
19
20
21
22
23
23
23
24
25
26
27
28
28
29
30

FIGURE 5.1. Grassland formations. 1. California steppe (*Stipa cernua–S. pulchra*), dense to medium dense, low to medium tall grassland. 2. Sagebrush steppe (*Artemesia tridentata–Agropyron spicatum–Festuca idahoensis*), dense to open grassland with dense to open shrub synusia. 3. Grama–needlegrass–wheatgrass prairie (*Bouteloua gracilis–Stipa comata–Agropyron smithii*), short, open to fairly dense grassland. 4. Wheatgrass–needlegrass prairie (*Agropyron smithii–Bouteloua gracilis–Stipa comata–Stipa viridula*), moderately dense, short or medium tall grassland. 5. Wheatgrass–bluestem–needlegrass (*Agropyron smithii–Andropogon gerardii–Schizachyrium scoparium–Stipa spartea–Bouteloua curtipendula*), dense, medium tall to tall grassland. 6. Tall-grass prairie (*Andropogon gerardii–Panicum virgatum–Sorghastrum nutans*), dense vegetation of tall grasses and many forbs. 7. Nebraska sandhills prairie (*Andropogon gerardii–A. hallii–Schizachyrium scoparium–Calamovilfa longifolia–Redfieldia flexuosa–Muhlenbergia pungens*), medium dense to open grassland, tall to medium tall, with inclusions of sand "blowouts." 9. Mixed prairie (*Schizachyrium scoparium–Bouteloua curtipendula–B. gracilis*), dense, medium tall grassland with many forbs. 10. Trans-Pecos shrub savanna (*Flourensia–Larrea–Bouteloua breviseta–B. trifida–Muhlenbergia porteri*), dense to scattered shrubs with short grasses, with inclusions of grama–tobosa prairie (*Bouteloua gracilis–Hilaria mutica*, short grasses) and oak–juniper woodland. 11. Grama–tobosa shrubsteppe (*Larrea divaricata–Bouteloua eriopoda–Hilaria mutica*), short grasses with very open to dense shrub synusia. 12. Mosaic of grama–tobosa shrubsteppe (formation 11) and pinyon–juniper woodland (scattered low evergreen trees with many grasses, including *Bouteloua curtipendula, B. gracilis, B. hirsuta, Oryzopsis hymenoides, Sporobolus cryptandrus*). 13. Mosaic of pinyon–juniper (see formation 11), coniferous forest, and Great Basin shrub formation. 14. Mosaic of pinyon–juniper, coniferous woodland, blackbrush, and galletta (*Hilaria jamesii*) shrubsteppe. 15. Grama–galletta steppe (*Bouteloua gracilis–Hilaria jamesii–Andropogon spp.–Bouteloua curtipendula–Oryzopsis hymenoides*), low to medium tall grassland with few woody plants. 16. Alpine meadows and barrens (*Poa–Festuca–Bromus–Phleum–Deschampsia*), medium tall, moist grass meadows, and short tundra dominated by broad-leaved forbs. 17. Mesquite–buffalograss (*Prosopis–Buchloë–Bouteloua gracilis–B. hirsuta*), short grasses with scattered low trees and shrubs and low needleleaf evergreen shrubs. 18. Juniper–oak savanna (*Schizachyrium scoparium–Quercus–Juniperus–Bouteloua* spp.–*Buchloë*), savanna with dense to very open synusia of broadleaf deciduous and evergreen low trees and shrubs. 19. Mesquite–acacia savanna (*Andropogon littoralis–Setaria macrostachya–Prosopis–Acacia*), dense to open grassland with broadleaf shrubs scattered singly or in groves. 20. Blackland prairie (*Schizachyrium scoparium–Stipa leucotricha–Bothriochloa saccharoides*), medium tall, moderately dense grassland. 21. Fayette prairie (*Schizachyrium scoparium–Buchloë–Bothriochloa–Paspalum*), medium tall, rather dense grassland with scattered open groves of broadleaf deciduous trees. 22. Coastal bluestem–sachuista prairie (*Bothriochloa–Spartina spartinae*), medium tall to tall, dense to open grasslands. 23. Mosaic of tall-grass prairie (formation 6) and oak–hickory forest. 24. Oak savanna (*Quercus–Andropogon*), tall-grass prairie with broadleaf deciduous trees scattered singly or in groves. 25. Cedar glades (*Quercus–Juniperus–Sporobolus–Andropogon*), low to medium tall open grassland with scattered evergreen shrubs and groves of broadleaf deciduous trees. 26. Blackbelt (*Liquidambar–Quercus–Juniperus–Sporobolus*), tall or medium tall broadleaf deciduous forest with concentrations of low needleleaf evergreen trees with patches of tall-grass prairie (formation 6). 27. Northern cordgrass prairie (*Spartina–Distichlis*), dense, medium tall grassland. 28. Live oak–sea oats (*Quercus–Uniola paniculata*), open grassland to dense shrubs and groves of low broadleaf evergreen trees. 29. Florida prairie (*Serenoa–Muhlenbergia–Aristida–Taxodium–Mariscus–Magnolia–Persea*), moist flats dominated by *Serenoa* with inclusions of hammocks and coniferous trees.

Mountains to Ohio (137) and from the south-central Canadian prairie provinces (23, 24, 86) into Mexico. Also, various grassland "islands" within the Rocky Mountain area, such as those on high plateaus or in arid valleys, added to the total North American land mass that shifted from forest to grassland in the Tertiary. If one also considers coastal grasslands and various types of savanna, North American grasslands included a vast diversity of vegetational types (76).

The prairie provinces of the grasslands have normally been divided into tall-grass (13, 147, 148) and short-grass prairie, corresponding in general to longitudinal precipitation patterns (147). A transitional zone between these two types is also usually recognized (1, 133, 139, 150, 151). Many authors believed that the area in which characteristic dominants of both of these grassland types grew had features that were more or less unique and termed this the "mixed-grass" prairie (1, 151).

The short-grass prairie is dominated as far north as Nebraska and southeastern Wyoming by two species—buffalograss (*Buchloë dactyloides*) and blue grama (*Bouteloua gracilis*). These grasses are relatively resistant to and may increase with moderate grazing pressure (13, 78, 136). Farther north (59, 106, 123) where buffalograss was rare, needlegrasses (*Stipa* spp.) and wheatgrasses (*Agropyron* spp.) were commonly associated with blue grama. Buffalograss spreads extensively by stolons; one plant may produce 650 ft of stolons in a single growing season (152). Blue grama (110), needlegrasses, and wheatgrasses are not stoloniferous.

Northern grasslands (23, 24, 86), at all longitudes, tended to be dominated by "cool-season" (pooid) grasses, whereas southern grasslands tended to be dominated by "warm-season" (panicoid and chloridoid) grasses (Table 5.1). The region in central Kansas and Nebraska where east–west and north–south tendencies intergrade is therefore the heartland of the North American prairie; the classic studies of this region by Weaver and associates in Kansas (147, 148) remain major milestones in definition of the prairie. Curiously, almost exactly at this heart of the grassland lies a large area (in west-central Nebraska) whose vegetation is shaped as much by its underlying sandy substrate as by the climate. These sand hills possess many special characteristics (134), some of which are shared by smaller sand areas in northeastern Colorado (109). The flora and fauna of sand prairies of eastern states such as Illinois (122) also differ from those of surrounding prairies on richer soils. Where the southern tall-grass prairie interfaced with the Texas coastal prairie, a "Blacklands Prairie" region had many unique features (19). The Palouse prairie (128) of the Northwest and other isolated grasslands of the mountain states (20, 164) possess unique characteristics, many of which are associated with high elevation. Finally, the so-called Desert Plains of western Texas, New Mexico, and Arizona, covered today with a copious overgrowth of mesquite (*Prosopis juliflora*), sage (*Artemisia tridentata*), and creosote bush (*Larrea divaricata*), were probably semiarid grasslands at the time of pre-Columbian

TABLE 5.1. Dominant grasses of the tall-grass prairie.[a]

	North	West-Central	East-Central	South
Pooideae				
Agropyron pseudorepens	nd	L	nd	nd
(Agropyron repens)[b]	H	H	H	H
Calamagrostis canadensis	L	L	L	L
Elymus canadensis	SL	L	nd	nd
Koeleria cristata	H	nd	S	nd
(Poa pratensis)	SL	nd	HS	nd
Stipa comata	SL	nd	HS	nd
Stipa spartea	HL	nd	HS	nd
Stipa leucotricha	nd	nd	nd	H
Chloridoideae				
Bouteloua curtipendula	H	HSL	H	H
Bouteloua gracilis	H	HS	nd	nd
Bouteloua hirsuta	nd	HS	nd	nd
Buchloë dactyloides	nd	H	nd	nd
Schedonnardus paniculatus	nd	S	nd	nd
Spartina pectinata	nd	nd	HL	nd
Sporobolus cryptandrus	nd	S	nd	nd
Sporobolus heterolepis	nd	S	H	nd
Panicoideae				
Andropogon gerardii	L	L	HSL	L
Bothriochloa saccharoides	nd	nd	nd	H
Dichanthelium scribnerianum	nd	HL	nd	nd
Dichanthelium oligosanthes	nd	nd	nd	S
Panicum virgatum	nd	H	SL	nd
Schizachyrium scoparium	H	HSL	H	H
Sorghastrum nutans	nd	HL	HSL	nd

[a]H, High prairie; S, sloping prairie; L, low prairie; nd, not dominant.
[b]Parentheses denote nonnative species.

colonization (14, 15, 156, 157). Actually, the Desert Plains consist of a wide variety of grassland associations, which depend on elevation, topography, substrate, and proximity to the Pacific Ocean and Gulf of Mexico (Fig. 5.2).

Even the driest North American plains were traversed by major river systems (i.e., Missouri, Arkansas, and Platte), and the associated floodplains supported localized intrusions of vegetation otherwise characteristic of more eastern longitudes. Even on the uplands, however, as moisture permitted, there was some tendency for occurrence of upland trees in pockets of tall-grass prairie (44, 106) and, to the east, forest islands (oak–hickory) of increasing size. Open grassland with isolated trees has usually been termed "savanna," whereas edges between forest islands and grassland have been thought of as ecotonal habitat. In any event, the summation of grassland–tree interface actually constituted a large habitat unit. In this chapter, for the sake of simplicity, we consider such habitat simply "savanna." The prairie pockets in the savanna had a substantial fraction of

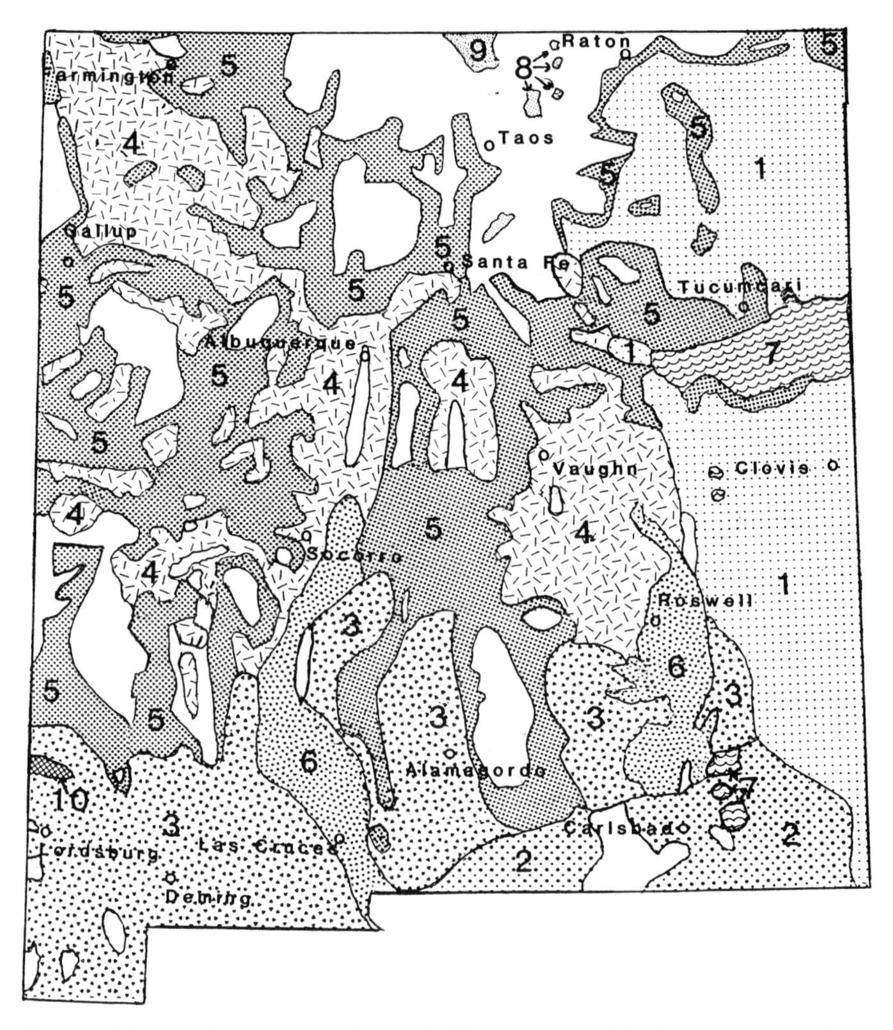

FIGURE 5.2. New Mexico grasslands. 1. Short-grass prairie (*Bouteloua gracilis–Buchloë dactyloides*), dense grassland of short grass, with significant inclusions of *Hilaria jamesii, Bouteloua curtipendula, Agropyron smithii,* and *Schizachyrium scoparium*. 2. Trans-Pecos shrub savanna (*Fluorensia–Larrea–Muhlenbergia porteri–Bouteloua* spp.–*Sporobolus* spp.), dense to scattered shrubs with short grass. 3. Grama–tobosa shrubsteppe (*Larrea divaricata–Bouteloua eriopoda–Hilaria mutica*), short grasses with very open to dense shrub synusia, with inclusions of *Muhlenbergia porteri, Sporobolus* spp., and *Aristida* spp. 4., Grama–galletta steppe (*Bouteloua gracilis–B. eriopoda–Stipa neomexicana–Hilaria jamesii–Schizachyrium scoparium–Bouteloua curtipendula–Oryzopsis hymenoides–Muhlenbergia torreyi–Sporobolus cryptandrus–S. airoides*), low to medium-tall grassland with few woody plants. 5. Pinyon–juniper grasslands (*Pinus–Juniperus–Bouteloua curtipendula–B. eriopoda–B. gracilis–B. hirsuta–Oryzopsis hymenoides–Sporobolus cryptandrus–Muhlenbergia* spp.), scattered low evergreen trees with many grasses. 6. Creosote bush–tarbush shrubsteppe (*Larrea–Fluorensia–Muhlenbergia porteri–Hilaria mutica–Bouteloua eriopoda–Scleropogon brevifolius*, fairly dense to open stands of shrubs and grasses, with significant inclusions of gyp flora. 7.

forbs (158). Cicaellids with life histories that involve alternate existence on herbaceous and woody hosts utilize both of these habitat types.

Habitat Turnover

Several major types of habitat turnover drastically disequilibrated the pre-Columbian vegetation patterns. Perhaps the most obvious form of turnover has been in the prairie, with the literal "turning over" of the sod. In place of the initial diverse mixture of grasses and forbs, a vast landscape has arisen that is dominated by monocultures. Many of these plants are annual grasses. In the drier prairie provinces, wheat (*Triticum aestivum*) has essentially become the new dominant. Other areas have been turned over to other small grains, such as barley (*Hordeum vulgare*) and oats (*Avena sativa*). Another grass [corn, or maize (*Zea mays*)] has become the principal dominant over large areas of the tall-grass prairie, and to the south, *Sorghum bicolor* has become a new dominant.

These familiar examples, however, represent only one aspect of the turnover of Nearctic grasslands. Particularly in the eastern states, but in all states to a significant degree, grassland has been turned over from native grasses to lawns. To some extent, these new grasslands are less different from the primordial grasslands than cultivated crops. Only the best kept of lawns are monocultures. Most often, despite the best efforts of the lawnkeeper, these grasslands tend more toward the natural scheme of intermixed grasses and forbs. Despite the massive merchandising of bluegrasses (*Poa* spp.) and fescues (*Festuca* spp.), there have been substantial efforts to use warm-season grasses (e.g., *Buchloë*) for lawns in the short-grass prairie regions and panicoid grasses such as St. Augustine grass in the southeast, in accord with natural vegetational patterns in those regions.

More subtle forms of habitat turnover have occurred in the short-grass prairie, with a shift in the dominant herbivore from *Bison* to *Bovis*. Although it has been correctly pointed out (124) that the prairie has probably been subjected to grazing pressures throughout modern geological times, these pressures have certainly intensified. In the Desert Plains especially, but also throughout all the plains states, dramatic changes in dominance hierarchies of grasses have occurred (14, 15, 69, 140, 156, 157). In areas where overgrazing has been especially severe, native grasses have been replaced entirely by less suitable introduced grass species or by forbs or shrubs (12) that possess more sophisticated defense against herbivores.

Shinnery (*Quercus–Schizachyrium scoparium*), low shrubs intermixed with grasses, including *Sporobolus* spp., *Bouteloua curtipendula* and *B. eriopoda*. 8. Alpine meadows and barrens (*Poa–Festuca–Bromus–Phleum*). 9. Montane grassland (*Agropyron–Stipa–Festuca*), medium to tall grassland. 10. Creosote bush–bur sage (*Larrea–Franseria*), open stands and shrubs and dwarf shrubs, with *Bouteloua eriopoda, Muhlenbergia porteri* and *Hilaria mutica*.

Although it is difficult to estimate the area of the plains in which overgrazing has occurred, it is probably not an overestimate to say "most of it." In areas such as Theodore Roosevelt National Park or Samuel Ordway Prairie (North Dakota) or Custer State Park (South Dakota), where bison herds have been reintroduced and are being managed thoughtfully to approximate the densities that existed at the time of pre-Columbian colonization, there may be some chance of reconstructing the grasslands as they once were. Also, well-managed ranches, including some in the Kansas Flint Hills, Nebraska Sand Hills, or Davis Mountains of Texas, may also offer significant opportunities for grassland preservation.

Because grassland species compositions and densities were determined by precipitation (77), significant turnover and some modification of biome boundaries occurred with the great drought in 1931–1934. Drought-susceptible plants experienced major reductions (135, 149) and competitive balances were altered (146), especially in association with grazing.

Other grassland turnover has been responsible for significant faunal changes. Within the last 30 years, an area of North America equivalent to the state of Ohio has been "paved." Actually, this figure, when given, includes the land incorporated into the massive interstate highway system, which was begun in the 1950s. A substantial fragment of the land is in fact devoted to new grassland. All too often, such highway plantings are monocultures of introduced grasses. In the north, tall fescue (*Festuca arundinacea*) has been employed; in the deep south, bahiagrass (*Paspalum notatum*) is commonly used.

Bahiagrass has also been utilized for other purposes. In some regions of Florida, bahiagrass plantings extend from horizon to horizon, consisting of lawn, roadside planting, and pastures. In fact, the conversion of eastern forest lands into pasture represents still another major grassland turnover. In the south, bermudagrass (*Cynodon dactylon*) is probably the major introduced pasture grass, whereas to the north, introduced pooid grasses such as timothy (*Phleum pratense*), fescues (*Festuca rubra,* etc.), and Kentucky bluegrass (*Poa pratensis*) dominate pasture and meadow habitats. The planting of bluegrass has been so intensive that even thoughtful students of pristine grasslands such as Carpenter (16) and Küchler (76) felt that *P. pratensis* had acceded to the state of a dominant on upland grassland habitats throughout the tall-grass prairie region, and would remain so even if "man were removed from the scene" (76). However, although *Poa pratensis* does dominate many areas of unplowed tall-grass prairie, especially from northeast Kansas north, it is apparently lack of fire that sustains bluegrass dominance (L. Hulbert, personal communication). For example, in the Flint Hills of Kansas, where ranchers have burned their grasslands frequently for many years, bluegrass is a minor component of grasslands or is absent. Also, in states to the east of Kansas, pastures that appear to be dominated by bluegrass become dominated instead by bluestems (*Andropogon* spp.) and indian grass (*Sorghastrum nutans*) when they are frequently burned.

The introduction of new grasses in western ranges has been intense in some areas. Among these might be listed weeping love grass (*Eragrostis curvula*), and kleingrass (*Panicum coloratum*), which have been planted extensively in Oklahoma and Texas, respectively. Weeping love grass has also been widely used as a roadside grass in the plains states and in the east. Crested wheatgrass (*Agropyron cristatum*) has been widely used in arid and semiarid regions to replace or supplement native perennial bunch grasses.

A significant aspect of habitat turnover involves the creation of grassland from forest. Perhaps half of the eastern deciduous forest has been converted into one or another of the various types of new grassland. Because in most cases some trees (in the form of hedgerows, ornamental plantings, etc.) remained, four centuries of European colonization have essentially witnessed the creation of a "neosavanna" in the eastern United States, which is dominated by (i) suburban grassland, (ii) crop monocultures, or (iii) pasture and meadow. The reality of this biome conversion is underscored by obvious eastward displacement of prairie birds such as Upland Sandpiper (*Bartramia longicauda*), Dickcissel (*Spiza americana*), Bobolink (*Dolichonyx oryzivorus*, and Eastern Meadowlark (*Sturnella magna*) into grasslands from Ohio to the eastern seaboard.

Inevitably, significant shifts in the species compositions of the grassland insect fauna have occurred with this massive habitat turnover. One of our main points will be to underscore the effect that these shifts have had on cicadellid populations. These events constitute the ecological context of the "appearance" of "pest" populations on the new species that man has chosen as new dominants of the North American grasslands.

Methods

Field Methods

Two basic approaches were used for collection of cicadellids from grasses and forbs. Earlier workers collected cicadellids with sweeping nets. Over the years, as more detailed records were taken with each collection, approximations of insect density were made by counting the numbers of sweeps taken and using that number as a rough estimate of the collecting effort. Deficiencies in this technique have been clearly recognized (29, 87). In general, plants that occur in dense stands are difficult to sweep efficiently. An even more serious problem involves insects that feed on roots or within the thatch of their grass hosts (10). Such insects can be overlooked in samples taken by sweeping. Collection of insects by vacuum (62) solves some of these problems; this is particularly true if, as in sweeping, each "catch" is monitored and separated by aspirating the insects immediately after the sweep. Since most genera and many species can be recognized in the net by an experienced collector, it is possible, by immediate monitoring, to assess the success or failure of individual collection attempts and to make improved decisions concerning the nature, intensity, and direction of further collection activities.

Insects collected in sweep or vacuum samples were aspirated from the net into vials of 70% ethyl alcohol and dried at a convenient later time. Collections that contained soft-bodied insects were passed through several changes of Cellosolve. Each collection was accessioned, and the cicadellids were sorted to species. Sorted, labeled series of each species were stored in gelatin capsules. Thus, all collected cicadellids were retained as vouchers. Representative specimens of many series were mounted and identified by conventional taxonomic procedures, which involved, in many instances, dissection of the male genitalia. Identification of dissected specimens followed published taxonomic revisions (6–9, 25, 30, 31, 34, 51, 53–57, 73–75, 79, 94, 95, 97, 98, 114, 118–121, 166–168). All specimens are currently stored in our collection at Beltsville.

Major difficulties were encountered in finding host stands of sufficient purity to enable collections to be made from them that were representative of the assemblage of insects appropriate to that host ("guild," in part, of some authors: 35–37, 82, 111). Unlike trees or shrubs, prairie grasses and forbs tend in many cases to grow in stands that are, by the very nature of prairie itself, mixed. It might even be correctly argued that for some hosts that never grow naturally in pure stands, attempts to define a specific faunal assemblage would be artificial. However, failure to ascertain host relationships has seriously misled many students of grassland cicadellids. For example, "grass-inhabiting" insects of some authors have turned out, in our studies, to be ascribable to forbs, or to *Carex* or *Juncus* species.

In many cases, pure collections could be made from disturbed locations, where the diversity of the initial vegetation had been substantially reduced. For example, early successional habitats often had reduced plant diversities, and accordingly higher numbers of pure host patches. Also, some grasses grow in restricted habitats, such as those with a considerable degree of alkalinity (58). In such cases, it was not uncommon to find pure, or nearly pure, stands of grasses such as alkali sacaton (*Sporobolus airoides*), salt grass (*Distichlis spicata* var. *stricta*), or alkali grass (*Puccinellia airoides*). Sandy regions were also characterized by pure patches of grasses [e.g., sand reed (*Calamovilfa longifolia*)].

A substantial fraction of our collections were made in plots of grass (or forbs) that had been planted by researchers, soil conservation workers, seed companies, or highway departments. Of these, the large collections in Plant Material Centers of the USDA Soil Conservation Service were very useful. Such collections enabled the species, variety, and time of planting to be ascertained. Particularly if such plantings were spared major disturbances, such as repeated pesticide applications, annual burning, or frequent mowing, it was not unusual to find a substantial subset of the appropriate cicadellid assemblage on the planted stand. This was, of course, especially true for plantings of native species that grew naturally in sufficient density in the surrounding region to provide colonizing propagules of the members of the appropriate cicadellid guild.

As a last resort, however, no method could replace the keen eye of the thoughtful collector in searching out relatively pure patches of native grasses or forbs in grassland preserves. It was in undisturbed natural areas that we usually found the largest assemblages of native fauna. Preserves maintained by The Nature Conservancy, because they are managed for maximization of their biota, were especially valuable. For field recognition of grasses and forbs in various grassland biomes, careful study of available literature on the flora of particular regions proved to be invaluable. Voucher specimens of host plants were preserved for later identification, and stored at the Beltsville laboratory.

Host Surveys

It was always desirable, and often absolutely necessary, to collect within a narrow time interval (7- to 10-day period) on a variety of native dominant grass and/or forb hosts in a given location (or region) to obtain host profiles for the major cicadellid species of the assemblage. Such surveys usually revealed broad patterns of host selection by generalists and contrasting specialization by monophagous or narrowly oligophagous species. Although this was best observed in surveys taken at regular intervals during an entire season, single surveys taken during a single time interval revealed valuable patterns representing distribution of species among hosts in the communities for a given season. Where only limited seasonal surveys were possible, we attempted to sample at the time of flowering or fruiting of the host, since seasonal cycles of specialists are often synchronized with that phase of their host's development. Our data from Maryland grasslands are especially robust in that they reflect collections throughout the season. Results from a survey such as that presented in Table 5.2, from a single survey at Beltsville in September 1974, therefore represent genuine and profound differences in cicadellid host selection.

TABLE 5.2. Cicadellid survey in Maryland.[a]

Strategy	Tribe or subfamily	Grass species	Cicadellid fauna
Perennial	Oryzoideae	*Leersia oryzoides*	*Graminella nigrifrons* 2F; *Draeculacephala portola* 15M, 10F; *D. mollipes* 3M; [*Paraphlepsius dentatus* 14M, 1F; *Graminella fitchii* 15M, 22F; *Chlorotettix* n. sp. 13M, 6F]
	Arundinoideae	*Danthonia spicata*	*Graminella nigrifrons* 2M, 2F; *Laevicephalus unicoloratus* 1M, 3F; *Endria inimica* 2M; [*Laevicephalus melsheimerii* 54M, 3F]

TABLE 5.2. *continued*

Strategy	Tribe or subfamily	Grass species	Cicadellid fauna
	Chloridoideae	*Aristida oligantha*	*Graminella nigrifrons* 1M, 3F; *Chlorotettix galbanatus* 3F; *Endria inimica* 1F; *Aceratagallia sanguinolenta* 2F; *Xestocephalus pulicarius* 1M; *Draeculacephala mollipes* 1M, 6F; [*Flexamia picta* 37M, 23F]
	Poeae	*Dactylis glomerata*	*Graminella nigrifrons* 65M, 55F; *Paraphlepsius irroratus* 2M, 1F; *Latalus sayi* 10M, 8F; *Polyamia weedi* 2M, 1F; *Stirellus bicolor* 10M, 9F; *Chlorotettix galbanatus* 2M, 1F; *Macrosteles fascifrons* 1F; *Exitianus exitiosus* 1M, 6F; *Endria inimica* 26M, 30F; *Psammotettix lividellus* 2F; *Planicephalus flavicostatus* 7M, 3F; *Balclutha hebe* 1M; *Forcipata loca* 2M; *Dikraneura angustata* 2M; *Aceratagallia sanguinolenta* 11M, 6F; *Agallia constricta* 13M, 6F; *Draeculacephala portola* 19M, 8F; *D. mollipes* 5M, 7F; *Graphocephala versuta* 2M; *Tylozygus bifidus* 1M, 1F; [*Xestocephalus pulicarius* 5M, 3F; *Amblysellus curtisii* 11M, 11F]
	Andropogoneae	*Andropogon virginicus*	*Graminella nigrifrons* 4M; *Draeculacephala mollipes* 5M, 2F; *Planicephalus flavicostatus* 2M, 2F; [*Stirellus bicolor* 32M, 40F; *Chlorotettix spatulatus* 1M, 2F; *C. viridius* 2M, 1F; *Flexamia sandersi* 8M, 4F; *Parabolocratus flavidus* 1M; *Polyamia caperata* 1M, 1F]
		Tripsacum dactyloides	*Graminella nigrifrons* 4M, 2F; [*Baldulus tripsaci* 58M, 39F; *Reventezonia lawsoni* 2F]
Annual	Panicoideae	*Digitaria sanguinalis*	*Graminella nigrifrons* 8M, 10F; *Exitianus exitiosus* 5M, 6F; *Endria inimica* 4M, 1F; *Planicephalus flavicostatus* 27M, 23F; *Psammotettix lividellus* 2M, 3F

TABLE 5.2. *continued*

Strategy	Tribe or subfamily	Grass species	Cicadellid fauna
	Panicoideae	*Panicum dichotomiflorum*	*Graminella nigrifrons* 17M, 16F; *Paraphlepsius irroratus* 1M; *Stirellus bicolor* 1F; *Chlorotettix galbanatus* 11M, 22F; *Balcultha impicta* 1F; *B. abdominalis* 1M, 6F; *Planicephalus flavicostatus* 1M, 1F; *Aceratagallia sanguinolenta* 1M, 1F; *Draeculacephala portola* 12M; *D. mollipes* 4M; *D. delongi* 8M; *Graphocephala versuta* 1M; *Tylozygus bifidus* 12M, 2F; *Plesiommata tripunctata* 1M

[a] Survey conducted on undisturbed pure grass stands at Beltsville Agricultural Research Center, September 3–11, 1974. One hundred sweeps were made on each host. The results demonstrate consistent differences between fauna of annual versus perennial grasses, and among different perennial grasses. Brackets indicate specialist on given host.

Laboratory Experiments

Although field studies were necessary to obtain information on the natural host range of cicadellids, determination of experimental host ranges offered an expanded and subtly different perspective (103). Experimental ranges were determined by growing grasses from seed in our greenhouse at Beltsville or by transplanting host plants from the field. Either method usually entailed a substantial period of adaptation before the growth characteristics of the stand were adequate to test the species' ability to support a cicadellid population. Insects were caged on the grasses by techniques commonly employed in plant disease vector research (154), and the caged colonies were held in environmental chambers provided with at least 1600 ftc of light, in most cases with a 16-hr photophase. Females collected from the field were pooled, and equal numbers were distributed at random to each of the tested grasses. Success in adapting insects to such caged plants, however, required a certain amount of judgment. Plants were changed at intervals of no more than 2 weeks and were then held for 2 weeks or more to observe hatching of eggs. However, plants that were considered to be unsuitable for cicadellid maintenance, for whatever reason, were changed, irrespective of the normal transfer intervals. Newly hatched nymphs were maintained, as were the adults, with regular changes of host plants. If a majority of adults survived for 5–7 days on a given host species, it was considered to be suitable for survival; if nymphs appeared, it was judged suitable for oviposition; and if the nymphs could be maintained on the host until they became adults and gave rise to a new generation of progeny, we considered that the cicadellid had been reared.

Patterns in Cicadellid Host Selection

General Patterns

Although many different general host selection patterns emerged from our studies, the patterns can be divided into six basic types.

(i) *Cicadellids confined to woody plants*. Although our study was designed (more or less) to exclude the study of leafhoppers confined to woody plant hosts, the study of grassland cicadellids necessarily encompasses the tribe Deltocephalini. It is of great interest, therefore, to define the general evolutionary patterns within this tribe. A number of deltocephaline leafhoppers, such as *Fieberiella florii,* appeared to be confined to woody plant hosts. All stages of such species, nymphs and adults, develop on woody plants.

(ii) Some species (e.g., *Paraphlepsius irroratus, Scaphytopius acutus, Colladonus clitellarius*) utilize both woody and nonwoody plants at different stages of their life cycle (48). Cicadellids with this strategy, which may enable them to transmit plant disease agents from one host species to another (49), are common in savanna or savannalike habitat. The diversity of cicadellids at woodland ecotones was especially high; many species found in this habitat were associated with either of a set of alternate hosts, but were not found on those hosts where they occurred in each other's absence. For example, nymphal stages of *Danbara aurata* fed on *Lactuca canadensis,* but moved to *Pinus* after molting to the adult stage.

(iii) Certain species appeared to be confined entirely to broad-leaved dicots. Within this stipulation, there appeared to be a wide range of degrees of specificity. For example, *Scaphytopius cinereus* was collected from several forb families in the Kansas prairie, confirming observations of Blocker and his colleagues (10, 26). Other species were confined to a single genus of forb (e.g., *Norvellina chenopodii;* hosts: *Chenopodium* spp.; *Texananus areolata:* hosts: *Helianthus maximilliani* or *H. mollis*).

(iv) A few species were found on both broad-leaved herbaceous plants or grasses. Probably the best known example is *Macrosteles fascifrons,* which has been found by others on many hosts (82) and can be routinely reared in insectaries on rye, aster, or a variety of other plants. Apparently, European species of *Macrosteles* are also dispersive generalists (107, 108, 142). *Agallia constricta* is another polyphagous cicadellid.

(v) Some species were found to have wide host ranges within the Poaceae. Such species usually had attracted the attention of other workers, and published records can be found that indicate wide natural host ranges. These insects include *Graminella nigrifrons, G.* (= *Deltocephalus*) *sonorus, Endria inimica, Exitianus exitiosus, Balclutha neglecta, B. hebe,* and several *Draeculacephala* species.

(vi) The vast majority of grassland leafhoppers exhibit considerable specificity. These range from the absolutely monophagous *Baldulus trip-*

saci and *Reventezonia lawsoni* (host: *Tripsacum dactyloides*) to certain specialists of pooid grasses, which may range within, but never beyond, the various tribes of that subfamily (e.g., *Psammotettix lividellus*), to specialists of panicoid grasses [*Chlorotettix spatulatus* (hosts: *Dichanthelium commutatum, Schizachyrium scoparium, Sorghastrum nutans*)].

Seasonal and Geographic Variation in Host Selection

Although remarkable consistency in host selection behavior was observed within prairie habitats, some seasonal and geographic variation was noted. Longitudinal variation was especially apparent west of the Mississippi, as aridity gradually increased. In general, with increasing aridity, there was an increasing tendency for "refugees" to be separated from their normal host plants during periods of seasonal or unseasonal adversity. Some grasses (e.g., *Distichlis stricta* in arid alkaline regions, *Dichanthelium scribnerianum* in sandhill regions, or *Cynodon dactylon* in semiarid regions of the southwest) appeared to serve as refugia for cicadellids displaced from other hosts. Although such "refugees" usually colonized the refugium at low densities, in some cases massive movement was evident (e.g., escape of *Diplocolenus configuratus* from mature annual *Bromus* species to less abundant warm season grasses such as *Redfieldia flexuosa* in sand hill habitats). The appearance of such refugees, however, has a fundamentally different interpretation than that of "accidentals" (individuals collected from plants that have no role in maintenance of the species). In contrast to host plant use by accidentals, use of alternate hosts by oligophagous cicadellid species may increase their fitness in arid or semiarid regions and may be of some importance as far east as the Missouri tallgrass prairie. In Maryland, however, we noted little evidence of refugees even at the end of the season in September.

Latitudinal variation in host selection behavior was also noted. This variation, which was especially prominent in genera of tropical origin (e.g., *Stirellus* and *Graminella*), may be associated with the necessity for phenological synchronization (unpublished data).

Experimental Host Ranges of Cicadellids

Experimental host ranges of cicadellids were determined by laboratory rearing experiments. For these tests we chose six leafhopper species that we believed, on the basis of field data, to have widely different host preferences. These were *Graminella villica,* a specialist on *Paspalum laeve* at northern latitudes; *Graminella oquaka,* a specialist on switchgrass (*Panicum virgatum*); *Polyamia apicata,* a specialist on *Dichanthelium lanuginosum* and perhaps related species; *Polyamia obtecta,* a common species associated with grasses in sandy old fields; *Baldulus tripsaci,* a strict specialist on *Tripsacum dactyloides;* and a generalist, *Planicephalus*

TABLE 5.3. Performance of cicadellid species confined on various grass species.[a]

Grass species		Performance of given cicadellid species[b]					
Subfamily	Species	G.v.	G.o.	P.a.	P.o.	B.t.	P.f.
Pooideae	*Festuca arundinacea*	S	nd	nd	nd	nd	nd
	Festuca abyssinica	nd	nd	R	R	nd	R
	Festuca spectabilis	nd	nd	R	nd	nd	nd
	Festuca sulcata	nd	nd	nd	nd	N	nd
	Festuca rubra	nd	N	nd	nd	nd	nd
	Poa compressa	nd	N	nd	nd	N	nd
	Poa nemoralis	nd	nd	nd	nd	R	nd
	Poa pratensis	N	nd	nd	nd	nd	nd
	Calamagrostis canescens	N	nd	nd	nd	N	N
	Calamagrostis epigejos	nd	N	N	nd	nd	nd
	Phalaris tuberosa	nd	N	nd	nd	nd	nd
	Phalaris arundinacea	N	nd	nd	R	nd	O
	Elymus paboanus	N	nd	nd	nd	nd	nd
	Elymus cinereus	nd	nd	nd	nd	nd	R
	Elymus triticoides	nd	nd	nd	R	nd	nd
	Agropyron caninum	N	nd	nd	nd	nd	nd
	Agropyron elongatum	nd	N	N	nd	nd	nd
	Stipa viridula	N	N	nd	nd	N	nd
Panicoideae	*Paspalum laeve*	R	N	N	R	nd	nd
	Paspalum paniculatum	R	N	nd	nd	N	nd
	Paspalum jurgensii	nd	N	nd	nd	nd	nd
	Paspalum wettsteinii	N	nd	N	N	nd	nd
	Paspalum nicorae	nd	nd	R	N	nd	nd
	Panicum decompositum	R	nd	nd	nd	nd	nd
	Panicum dichotomiflorum	R	nd	nd	nd	nd	nd
	Panicum obtusum	R	nd	nd	N	nd	nd
	Panicum virgatum	nd	R	N	nd	nd	R
	Panicum bisulcatum	nd	N	nd	nd	nd	nd
	Panicum maximum	nd	N	nd	nd	N	nd
	Panicum clandestinum	nd	N	nd	nd	nd	R
	Tripsacum dactyloides	nd	nd	nd	nd	nd	R
	Brachiaria ramosa	R	R	R	R	nd	R
Chloridoidae	*Sporobolus wrightii*	nd	nd	R	nd	nd	nd
	Sporobolus airoides	nd	nd	nd	nd	nd	R
	Sporobolus usitatus	nd	N	nd	nd	nd	nd
	Muhlenbergia wrightii	nd	N	N	R	nd	nd
	Aristida sp.	nd	N	N	R	nd	nd

[a] R, Reared; O, oviposition; N, negative; S, survived; nd, not done.
[b] Cicadellid species: G.v., *Graminella villica,* a specialist on *Paspalum laeve* at northern latitudes. G.o., *Graminella oquaka,* a specialist on switchgrass *(Panicum virgatum).* P.a., *Polyamia apicata,* a specialist on *Panicum lanuginosum.* P.o., *Polyamia obtecta,* a habitat specialist in sandy old fields. B.t., *Baldulus tripsaci,* a specialist on *Tripsacum dactyloides.* P.f., *Planicephalus flavicostatus,* a generalist.

flavicostatus, that occurs on various grasses in several subfamilies. These cicadellids were collected in the field at the time of maturation of the first generation (June), and confined on various hosts in the laboratory. These grass species, mostly from the seed collection of the National Plant Material Center of the USDA Soil Conservation Service, had been chosen for their possible significance in conservation plantings in various parts of the United States.

The results (Table 5.3) demonstrated a great diversity of host selection behavior. Although two of the specialists showed their monophagous patterns clearly, one of them, *G. oquaka,* could be reared on a *Brachiaria* species, *B. ramosa*. These patterns were in sharp contrast to that of *Polyamia apicata* (a cicadellid that appeared in field studies to be a strict specialist), which could be reared on a variety of panicoid and nonpanicoid grasses. In this respect, it was little different from *Polyamia obtecta,* which in field studies appeared to be a habitat specialist (although it is possible that there may be an important undiscovered preference in its oviposition behavior). Finally, the widest spectrum of rearing was observed for the generalist *Planicephalus flavicostatus,* in accordance with field observations.

Our results are in general accordance with experimental host range studies performed on other heterometabolous and holometabolous insect species. It is unusual to observe experimental host ranges as narrow as those observed in the field. Many of the constraints that affect insect populations in the field are absent in the artifical circumstances established in the laboratory. In this setting the insects need not disperse to find an unusual host; they need only decide whether they wish to survive and/or breed on it. Drought conditions or other adversities may actually present them with similar choices in the field. In regions where occurrence of suitable plants of the main host species is unpredictable, it would seem highly adaptive for insects to develop a capacity for subsistence for at least short periods on unusual and perhaps suboptimal food plants. Although we successfully reared *P. apicata* on nonpanicoid grasses for one generation, it is not clear that sustained rearing would have succeeded. Even if it had done so, the fitness of the reared population may well be less than that of the field population.

Cicadellid Components of Grassland Guilds

We now discuss the cicadellid components of some of the "guilds" that are associated with North American grasses and forbs. We use this term in the sense intended in recent ecological literature (35–37, 111). Ross (112) commented at some length on coexistence of cicadellid species on specific hosts. Recent workers have stressed the diversity of community interactions, and have endorsed the general concept only with certain reservations (see following chapter). Almost all of the plant species discussed in this section are in fact attacked by several taxa of sap-sucking insects. We usually did not collect or identify the other auchenorrhynchous insects on these plants. For example, the set of xylem-feeding insects, including cicadelline leafhoppers, cercopids, and some fulgoroids, may in some instances stress the plant more than the phloem-feeding cicadellids. We agree with Denno and his associates (35–37) that a detailed study of the community ecology of grass or forb species should include at least the entire set of phloem-feeding insects.

Cicadellids of Native Dominant Perennials

One of the largest grassland cicadellid assemblages is confined to andropogonoid hosts. Species of the genus *Flexamia* are the most important components of these guilds (Table 5.4). The *Flexamia* species group, which includes *F. graminea, F. prairiana,* and *F. clayi,* apparently speciated without host transfer. Such speciation is most likely to have been allopatric. Outlying species of *Flexamia* collected from Illinois and Wisconsin (118) and Georgia (57) may in fact be geographic extremes of a cline of the *F. sandersi* complex. Other species also occur regularly on andropogonoid species, some of them over a wide geographic range. These include *Laevicephalus unicoloratus, Parabolocratus flavidus, Stirellus bicolor,* and *Chlorotettix spatulatus.* In the eastern states, especially Maryland, *C. spatulatus* is found predominantly on *Andropogon virginicus,* but in the tall-grass prairie, this species occurs on a number of forbs late in the season, when andropogonoid hosts may be unsuitable.

The *Panicum virgatum* guild has also been discussed earlier. This guild, which represents the fauna of one of the most widely distributed grass species of the Nearctic grasslands, also varies geographically to a wide degree (Table 5.5).

A large assortment of cicadellids of particular interest occur on *Sporobolus* species. Collections from *S. cryptandrus* in Nebraska suggest that *Unoka ornata, Dicyphonia ornata,* and *Athysanella curtipennis,* are regular inhabitants of this species. In Montana and New Mexico, *Unoka ornata* was found on the large dropseed alkali sacaton (*Sporobolus airoides*), and in New Mexico and Texas it was on the dimunitive *S. nealleyi.* In the southeast and east-central grasslands, *Sporobolus vaginiflorus* is the host of *Athysanella balli* and *Lonatura catalina.* Other dominant species of *Sporobolus,* such as *S. asper, S. heterolepis,* and *S. wrightii,* seem almost certain to have interesting cicadellid faunas as well.

Another chloridoid grass of the west that we have sampled is *Distichlis spicata* var. *stricta.* This proved to be one of the easiest grasses to find in pure stands, since it is adapted to levels of alkalinity, or salts, that are not tolerated by other plant species. A rich cicadellid fauna occurs on this grass, but was found to vary from site to site. Two *Lonatura* species were collected on this host, as well as *Athysanella kadokana.* It is possible that the propensity of this grass to grow in isolated locations may have led to the adaptation of a number of endemic species. Further collecting on this grass throughout arid regions of the west should be of particular interest.

Large numbers of cicadellids, of many species, were always found on *Bouteloua gracilis* and *Buchloë dactyloides. Flexamia curvata* appeared

TABLE 5.4. Cicadellids on andropogonoid grasses in different geographic locations.

Species	States in which cicadellid species occurs
On *Schizachyrium scoparium*	
Flexamia graminea	Missouri, Oklahoma, Kansas, Texas
Flexamia prairiana	Missouri, Kansas
Flexamia dakota	Oklahoma, Wyoming, North Dakota, Nebraska
Flexamia albida	Oklahoma, Iowa, Kansas
Stirellus bicolor	Oklahoma, Kansas, Nebraska
Laevicephalus unicoloratus	Missouri, Wyoming, Kansas, North Dakota, Nebraska
Chlorotettix spatulatus	Missouri, Kansas
Parabolacratus flavidus	Kansas
Paraphlepsius altus	Kansas, Nebraska
Athysanella incongrua	Oklahoma, North Dakota
On *Andropogon virginicus*	
Flexamia graminea	Missouri
Flexamia clayi	Maryland, Indiana, Ohio, West Virginia
Flexamia sandersi	Maryland, Virginia, Georgia
Stirellus bicolor	Maryland, Virginia, Missouri, Indiana, Ohio, West Virginia
Laevicephalus unicoloratus	Maryland, Virginia, Indiana, Ohio, West Virginia
Chlorotettix spatulatus	Maryland, Virginia, Missouri, Indiana
Parabolocratus flavidus	Maryland, Virginia, Missouri
Polyamia caperata	Maryland, Virginia
On *Andropogon gerardii*	
Flexamia prairiana	Kentucky
Flexamia atlantica	Kansas, South Dakota
Stirellus bicolor	Kansas, Kentucky
Laevicephalus unicoloratus	Maryland, Virginia, Indiana, Ohio, West Virginia
Chlorotettix spatulatus	South Dakota
Polyamia caperata	Kentucky, Illinois
On *Andropogon hallii*	
Laevicephalus vannus	Colorado, Nebraska
Chlorotettix spatulatus	Colorado
On *Bothriochloa* spp.	
Flexamia prairiana	Texas, Oklahoma, Kansas
Graminella mohri	Texas
Laevicephalus vannus	Texas
Chlorotettix spatulatus	Kansas
Athysanella texana	Texas
On *Sorghastrum nutans*	
Graminella mohri	Kansas, Nebraska, Minnesota
Stirellus bicolor	Virginia, West Virginia
Chlorotettix spatulatus	Maryland, Virginia
Polyamia caperata	West Virginia

TABLE 5.5. Cicadellid specialists of *Panicum virgatum* in different geographic locations.[a]

Region or locality	*Graminella* spp.			*Chlorotettix* spp.		*Flexamia* sp.	*Balclutha* spp.	
	oquaka	*mohri*	*aureovittata*	*fallax*	*spatulatus*	*atlantica*	*abdominalis*	*hebe*
Maryland	+	0	0	+	+	0	+	0
Chicago	+	0	+	0	+	0	+	0
Kansas	0	+	0	0	+	+	+	+
Texas	0	+	0	0	+	+	+	+

[a] +, Regular guild membership; 0, absent or highly irregular guild membership.

to be restricted to *Buchloë,* whereas *Flexamia flexulosa* and *F. abbreviata* appeared to be restricted to *Bouteloua*. A number of other leafhopper species occur in this association. However, because the cicadellid assemblages of these two grasses are large and complex, and because exceptional care is required to find situations in which the species grow independently, delineation of their ecology must await further study.

Guilds on pooid grasses tend to overlap extensively. Four main groups of species are involved. One group consists essentially of cicadellids (e.g., *Exitianus exitiosus* and *Planicephalus flavicostatus*) that, later in the season, move to annual chloridoid and panicoid grasses. A second group of species (e.g., *Psammotettix, Quontus,* and *Latalus* species) appears to be restricted to or to greatly prefer grasses in the tribe Poeae (e.g., *Poa, Festuca, Bromus,* and *Dactylis* species). A third group of species, apparently Eurasian in origin (e.g., *Doratura stylata* and *Athysanus argentarius*), contains species that are univoltine, but that cross widely among pooid grasses. Because they are univoltine, their biologies are tied to pooid perennials, and they do not utilize annual grasses later in the season. Rather, they estivate. A fourth group is restricted to narrow species groups within Poeae. This group includes species restricted to the *Agropyron–Elymus* complex (e.g., *Commellus comma, Dorycephalus platyrhynchus*), the Stipeae (*Commellus colon*), or *Glyceria* (*Chlorotettix fumidus*). Overlapping of cicadellid host ranges within pooid grasses explains, to a large extent, why some workers in Eurasia or in northern latitudes of North America noted little evidence of restricted host selection behavior (84,

TABLE 5.6. Cicadellid specialists of grasses of the Arundinoideae, Oryzoideae, and Bambusoideae.

Grass species	Region	Cicadellid fauna
Danthonia spicata	Maryland	*Laevicephalus melsheimerii*
	Illinois	*Laevicephalus melsheimerii*
Leersia oryzoides	Maryland	*Graminella fitchii, Chlorotettix* n. sp., *Paraphlepsius dentatus*
Leersia virginica	Maryland	*Graminella fitchii*
Arundinaria spp.	Southeast	*Arundanus* spp.

85, 141–143), and instead emphasized grassland structure (2, 84, 85) as a major determinant of cicadellid occurrence and distribution.

Cicadellids that specialize on one of the few representatives of arundinoid, oryzoid, or bambusoid grasses are of special interest, because their hosts are widely separated, taxonomically and phylogenetically, from the three dominant groups of grasses. As one might expect, *Leersia, Danthonia,* and *Arundinaria* have a unique cicadellid fauna (Table 5.6).

Cicadellid Specialists of Native Dominant Forbs

A substantial number of grassland cicadellids are associated with perennial forbs. These deep-rooted plants provide resources throughout the growing season. The forb guilds include both specialists and generalists. Among the generalists are *Macrosteles fascifrons,* a phloem-feeding supertramp, and *Cuerna* spp., xylem-feeding generalists of dry grasslands. Other cicadellid genera associated with forbs include *Scaphytopius, Mesamia, Norvellina, Driotura, Texananus,* and *Xerophloea.* Composite forbs have a rich biota; in particular, *Solidago* and *Helianthus* spp. have several specialists associated with them in obligate host relationships. In Table 5.7,

TABLE 5.7. Representative cicadellid inhabitants of prairie forbs.[a]

Forb species	Prairie		Cicadellid species[c]	Status[d]
	Type[a]	Region[b]		
Amorpha canescens	U	T	*Scaphytopius cinereus*	Ge
Helianthus annuus	U	T	*Mesamia nigridorsum*	Sp
Helianthus maximiliani	L	T	*Mesamia* spp.	Sp?
			Texananus areolatus	Sp
Aster ericoides	U	T	*Driotura robusta*	Sp
Erigeron spp.			*D. robusta*	Sp
Solidago spp.	UL	T	*Paraphlepsius solidaginis*	Sp
			Neocoelidia tumidifrons	Sp
			Driotura gammaroides	Sp
			Xerophloea peltata	Ge
			Graphocephala hieroglyphica	Ge
			Scaphytopius acutus	Ge
			Cuerna spp.	Ge
Petalostemom purpureum	U	T	*Ponana puncticollis*	Sp?
			Scaphytopius cinereus	
Gutierrezia spp.	U	S	*Driotura vittata*	Sp?
Artemisia ludoviciana	U	S	*Mesamia coloradensis*	Sp
Artemisia spp.	U	S	*Thamnotettix* sp.	Sp?
Chenopodium spp.	U	T	*Norvellina chenopodii*	Sp
Kuhnia	U	T	*Scaphytopius* spp.	Ge
Salvia pitcheri	U	T	*Scaphytopius acutus*	Ge
Echinacea pallida	U	T	*Scaphytopius cinereus*	Ge

[a] U, Upland; L, lowland.
[b] T, Tall-grass prairie; S, short-grass prairie.
[c] The generalist *Macrosteles fascifrons* was found on many prairie forbs.
[d] Sp, Specialist; Ge, generalist.

we list some cicadellid species that we have found to be regularly associated with prairie forbs.

Cicadellid Guilds on Introduced Grasses

Bermudagrass (*Cynodon dactylon*). Bermudagrass is perhaps the most important introduced perennial chloridoid grass in North America. The planting of this grass throughout the southern United States represents a huge natural experiment that affords a major insight into patterns of "crossing-over" of indigenous arthropod faunal elements. Our surveys on *Cynodon* (Table 5.8) demonstrated the absence of any continental "consensus" fauna. Rather, in each region, the *Cynodon* fauna consisted largely of generalist grass-inhabiting species or specialists (in varying degrees) on chloridoid grasses. Several species of *Flexamia* crossed over to this host. In the southeast, *F. producta* predominated, and in the west, *F. atlantica* and *F. inflata* were most common. Several cicadelline species, including *Carneocephala* spp. and *Ciminius* spp. occur on bermudagrass in southern and southwestern regions. Cicadellids that crossed over to *Cynodon* from other chloridoid grasses included *Polyamia weedi, Balclutha neglecta,* and *Chlorotettix viridius. Laevicephalus unicoloratus* crosses over to *Cynodon* from andropogonoid grasses. Other major components of *Cynodon* assemblages included common generalists (Table 5.8). Nielson

TABLE 5.8. Species assemblages of cicadellids on bermudagrass (*Cynodon dactylon*) in various geographic locations.[a]

Species	Florida	North Carolina	Maryland	Kansas	Nebraska	Oklahoma	Texas
Flexamia producta	0	+	+	0	0	0	0
Flexamia inflata	0	0	0	+	0	+	+
Flexamia atlantica	0	0	0	0	+	+	+
Polyamia weedi	0	+	+	0	0	0	0
Chlorotettix viridius	0	+	+	0	+	0	+
Chlorotettix spatulatus	0	0	0	+	+	0	+
Laevicephalus unicoloratus	0	0	+	0	0	0	0
Exitianus exitiosus	+	+	+	+	+	+	+
Graminella nigrifrons	0	+	+	0	+	0	+
Graminella sonorus	0	0	0	0	+	0	+
Graminella villica	+	0	0	0	0	0	0
Planicephalus flavicostatus	0	0	+	0	0	0	0
Draeculacephala spp.	+	+	+	+	+	+	+
Ciminius spp.	0	0	0	0	+	+	+
Carneocephala spp.	+	+	+	0	+	+	+
Balclutha guajanae	0	0	0	+	0	0	0
Balclutha neglecta	0	0	0	+	0	+	0
Athysanella occidentalis	0	0	0	0	+	0	+
Stirellus bicolor	+	+	+	0	+	0	+

[a] +, present; 0, not present in our surveys.

and Toles (96) studied patterns of distribution and isolation among species of *Carneocephala* on *Cynodon* in the southwestern desert. The occurrence of different, new *Carneocephala* species on some of these *Cynodon* islands suggested that speciation might have occurred within the few centuries since the introduction of this grass into the southwest. However, the simplest interpretation of such observations could involve host transfer of rare or geographically isolated *Carneocephala* species to *Cynodon* as it became established.

Weeping love grass (*Eragrostis curvula*). Weeping love grass has been introduced as a major range grass in Oklahoma. In Maryland and elsewhere this grass has been planted along highway right-of-ways. At Beltsville and elsewhere in the east it has been established on upland slopes by the Soil Conservation Service. Such plantings have provided an opportunity to assess the assemblage of crossovers onto this grass. In Oklahoma, *Flexamia atlantica* crosses over to weeping love grass. Throughout much of its range, however, the principal cicadellid crossovers onto this species are *Flexamia inflata* and *Laevicephalus unicoloratus*.

Kleingrass (*Panicum coloratum*). Kleingrass has been introduced as a range grass in Texas. On that host we found an assemblage of cicadellids that included *Spangbergiella vulnerata* and *Polyamia yavapai*. We do not know at present the native grasses that serve as hosts for these species.

Bahiagrass (*Paspalum notatum*). Although growth characteristics of bahiagrass in Florida, Georgia, and other southern states appeared to be excellent, this grass supported relatively low densities of cicadellids. The species richness of the assemblage also was low. One predominant specialist, *Graminella villica*, which has been collected on several other *Paspalum* species, was found on this grass. Other colonists of bahiagrass were in fact generalists that were also collected on a number of other grasses. These generalists included *Graminella nigrifrons*, *G. sonorus*, *Exitianus exitiosus*, and several *Draeculacephala* species.

Cicadellid Assemblages on Annual Grasses

Perennial but not annual hosts accumulate specialist cicadellid faunas. This is intuitively obvious from an understanding of the "escape" strategies of insects and their close interrelationship with phenology (39). Few, if any, annuals provide adequate resources for sap-sucking insects through the entire seasonal cycle. This is not to say that during the logarithmic growth phase of an annual plant that it may not be an extremely suitable host. Almost all of the host plants upon which cicadellids are reared in laboratories, for example, are annuals. In field situations where annual vegetation constitutes a significant amount of total standing biomass, highly dispersive cicadellid species usually manage to find and exploit the annuals.

In Table 5.2 we present data from a survey that was conducted in September 1974. Six perennial and two annual grasses, *Digitaria sanguinalis* and *Panicum dichotomiflorum,* were surveyed. The assemblages on these grasses showed considerable overlap, especially with regard to the common generalists (*Graminella nigrifrons, Exitianus exitiosus,* and *Planicephalus flavicostatus*). Perennials, especially chloridoid and panicoid grasses, had specialist faunas. On the annual grasses we found only a set of generalist or "fugitive species." Some of these highly dispersive generalists show general preferences within subfamilies. *Endria inimica,* for example, is more apt to colonize pooid and chloridoid grasses and is not especially conspicuous on panicoid grasses. Such a preference is not surprising, since *Endria,* which has an egg diapause, probably evolved as a generalist in the central prairie, in a region dominated by pooid and chloridoid grasses (21, 22). Conversely, generalists of the genus *Graminella* and *Exitianus* are more southern in origin, overwinter as adults, and lack escape mechanisms that would permit winter survival at northern latitudes. They have thus evolved in a region of the Nearctic where panicoid and chloridoid grasses co-occur. Even in regions where all of these generalist cicadellids occur together, and, to some extent, where they occur together on the same host species, the preferences that were adaptive in their region of origin are still evident.

Colonization of Poaceous Crops

We are now able to examine the colonization of gramineous crops as part of a process of colonization of Poaceae as a whole. Pest species are sometimes thought of as "appearing" as if they had been theretofore absent or at the very least inapparent. However, the cicadellids of poaceous crops are simply those faunal elements that were no doubt adapted, long before the North American grasslands were turned over to crops, for native annual grasses. Thus, the cicadellid fauna of corn (41–43) and wheat in temperate latitudes, for example, represents the more general colonists of andropogonoid grasses and, in particular, dispersive insects that regularly exploit native annuals. In the natural range of *Zea* in the Neotropics, several cicadellid specialists (*Dalbulus* spp.) colonize stands of the grass (88, 89, 92, 138). In the United States, several cosmopolitan species with especially wide host ranges that encompass several grass subfamilies (e.g., *Graminella nigrifrons* and *Exitianus exitiosus*) occur on maize, and may be responsible for transmission of destructive viral and/or mycoplasmal plant disease agents. From these examples, it is apparent that the understanding of the ecological and evolutionary context of pest "appearance" is surely a vital key in formulating rational management strategies for pest species.

Evolution of Cicadellid Host Selection

Role of Geological History

Although the earliest fossil insects (66) date to the Devonian period of the Paleozoic about 350 million years ago (130), the first evidence of the Hemiptera is from the Permian (Fig. 5.3). The earliest hemipterans were probably predaceous or fungivorous (127). Plant feeding must have evolved independently several times in the predominantly predaceous Heteroptera (127). Divergence of Auchenorrhyncha and Sternorrhyncha from the heteropteran stem may have occurred in the Permian (45, 71, 125). Tracing the early divergence of the stems of the current groups, including cicadellids, is made difficult by a paucity of fossils, the nature of the fossil material (which tends to consist only of wings), and difficulties in defining monophyletic groups in such taxa as Cercopoidea (66). Nevertheless, the weight of the evidence suggests that the Auchenorrhyncha and Sternorrhyncha had diverged by the late Permian, and that by the Upper Triassic period of the Mesozoic, the Cicadoidea (ancestral cicadas), Cicadelloidea (ancestral leafhoppers), and Fulgoroidea (ancestral fulgoroids) had separated (66). Conditions that prevailed at that time (Fig. 5.3) included a shift to gymnosperm rather than pteridosperm dominance that had begun in the Permian and the presence of primitive flowers that were probably insect-pollinated (125). Most of the modern insect orders were represented.

The cicadellid fauna of world grasslands must be of considerably more recent origin than the ancient stem from which it is derived, because grasses themselves date only to the early Tertiary, and even angiosperms to the early Cretaceous period (Fig. 5.3). Some cicadellid taxa (such as genera of Deltocephalini) may date to the Cretaceous when the world flora began to shift from gymnosperm to angiosperm dominance.

The North American grasslands, created by geological events of the Tertiary, are especially recent, in that the successive glaciations of the Pleistocene had an obvious major impact on the contemporary species assemblages (113). The glaciations had several major effects on speciation and extinction. These include: (i) decimation of regional biotas by the ice masses, followed by recolonization after glacial recession; (ii) radical modification of terrain at the interface between glaciated and nonglaciated regions; (iii) extensive shifts in biota of nonglaciated regions during changing climatic circumstances; (iv) drastic shrinkage of some biomes; and (v) fragmentation (by glacial advances) of populations of widespread species into subpopulations prone to allopatric speciation.

These effects might be expected (i) to have reduced the diversity of the northern plains fauna by causing extinctions of many taxa; (ii) to have produced unique faunal elements in regions such as the Nebraska sand hills, which were produced by wind action at the glacial interface; and (iii) to have altered regional faunas as major vegetation cover underwent

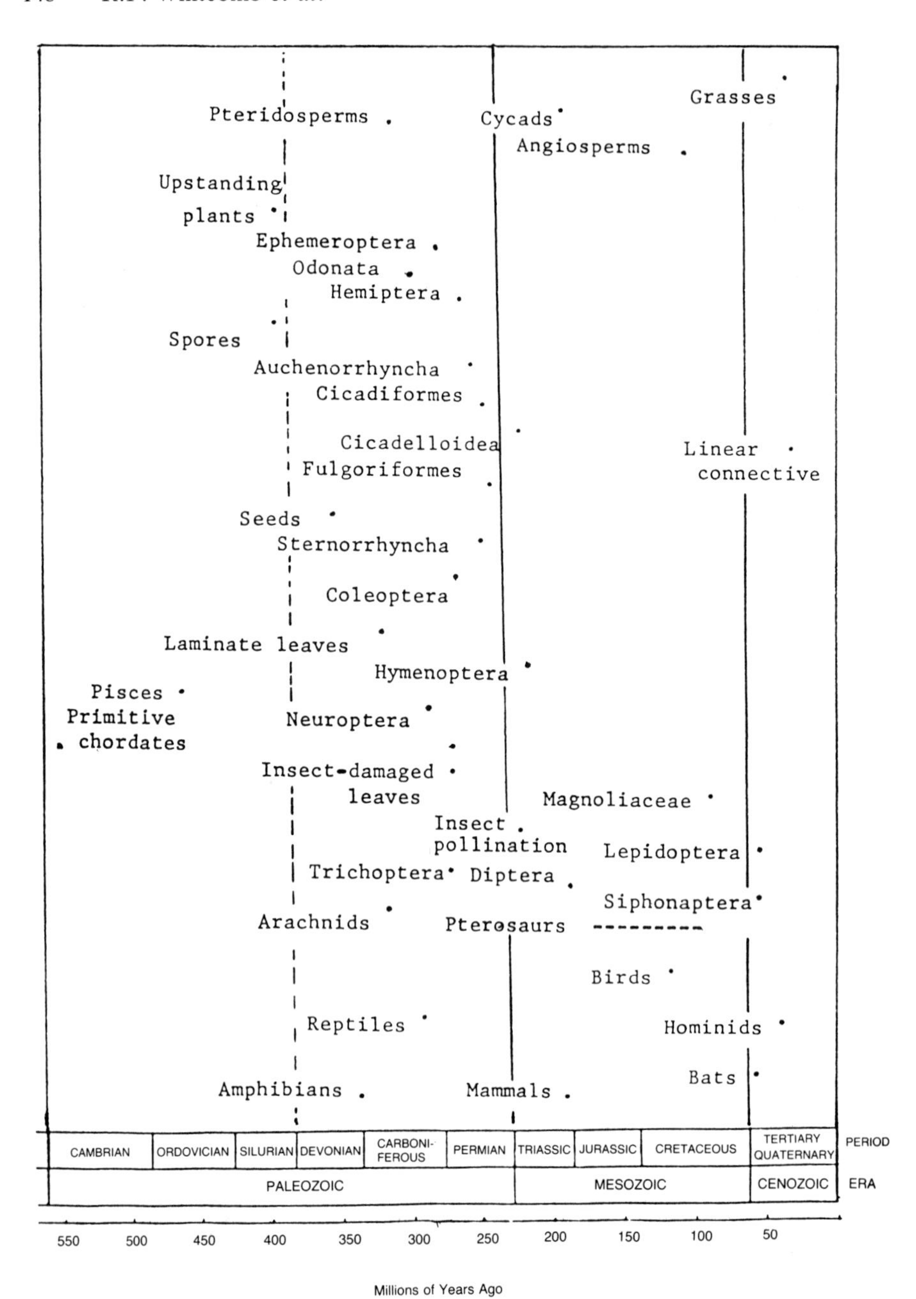

FIGURE 5.3. Important events in coevolution of insects and plants. Dots indicate approximate geological time of origin. [Modified from Smart and Hughes (125).]

complete turnover. For example, at the time of the Wisconsinan maximum, Iowa was covered by spruce forests; these gave way to deciduous forests, followed by prairie and savanna (165). Deciduous forest was replaced by prairie as far east as Ohio and Kentucky as the effects of glaciations diminished. Further, climatic changes in the Tertiary apparently caused the turnover of mesic grasslands into the semiarid Chihuahuan and Sonoran deserts (3, 72). Many elements of grassland biota in such deserts may thus be restricted today to relict grasslands at higher elevation. At the time of the Wisconsinan maximum the prairie may have been trapped by cool climates on its north and by the subtropical climates of Mexico on its south, and may therefore have occupied a relatively small area in south Texas and northern Chihuahua (115). Such a drastic shrinkage and geographic displacement would have been certain to have drastic effects on the species richness of its fauna, but would open the possibilities of speciation by relict populations under selective pressures of desertification that this region experienced after the glaciations. Even at the glacial maxima, a large region of contemporary northern and mixed-grass prairie, including the Black Hills of South Dakota (61) extending as far north as North Dakota, escaped glaciation. Although this area must have been largely covered by coniferous forest during the glacial maxima (80), it is possible that grassland species such as *Schizachyrium scoparium* may have remained on steep slopes, as in contemporary coniferous forests of the eastern Rockies. Particularly under selective pressure from a rapidly changing climate, cicadellid populations in such a region might quickly diverge from their geographically isolated counterparts, resulting in allopatric speciation. In summary, current distributions (67) of all prairie taxa undoubtedly reflect Pleistocene events (83).

Just as understanding the distribution and composition of contemporary species assemblages of North American grasslands requires a thorough understanding of the historical events of the Pleistocene, understanding dispersal of genera of Deltocephalini that feed on woody tissue may require historical reconstruction of divergences and dispersals on the more ancient time scale that take continental drift into account.

Grass Phylogeny

Our understanding of cicadellid host ranges was greatly enhanced by the realization that most cicadellid species were not confined to a single plant species, but that each leafhopper species has a unique host spectrum that corresponds (in most cases) to natural relationships among grass species. It was therefore necessary for us to consider grass relationships. In consultation with F.W. Gould, we prepared a two-dimensional representation of general grass relationships (Fig. 5.4). The diagram reflects, essentially, the views of classification expressed by Gould and Shaw (50). Although exact representations of grass phylogeny are not available, it was felt that

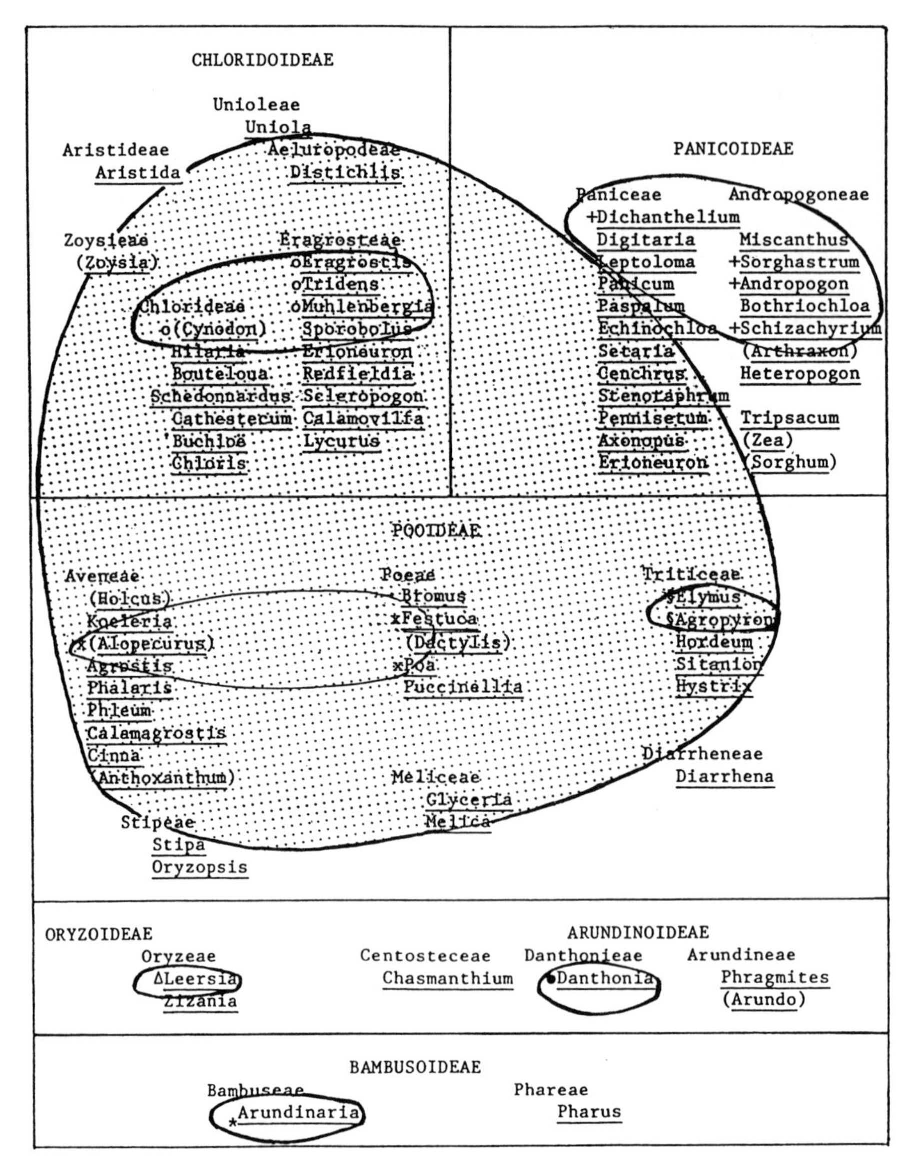

FIGURE 5.4. Host spectra of various deltocephaline leafhoppers interpreted on a two-dimensional diagram [from concepts of Gould and Shaw (50)] of relationships among grass genera. Shaded area: host spectrum of *Endria inimica*. Other symbols represent spectra of (○) *Polyamia weedi*, (+) *Chlorotettix spatulatus*, (×) *Psammotettix lividellus*, (§) *Dorycephalus platyrhynchus* and *Commellus comma*, (△) *Graminella fitchii*, *Paraphlesius dentatus*, and *Chlorotettix* sp. (•) *Laevicephalus melsheimerii*, and (*) *Arundanus* spp.

that bambusoid grasses (e.g., *Arundinaria*) were the most phylogenetically remote of the Nearctic grass subfamilies. Oryzoid grasses (e.g., *Leersia*) were also felt to be remote, although perhaps not so much so as the arundinoids (e.g., *Phragmites, Danthonia*). Three subfamilies, the pooids, chloridoids, and panicoids, which contain the vast majority of Nearctic grass genera (50) were felt to be more closely related to one another than to the other subfamilies.

These general concepts, which express a qualitative sense of grass relationships, can be diagrammed (Fig. 5.4) to provide a preliminary means of "mapping" cicadellid host ranges. Although it would have been far more satisfactory if an objective phylogeny of grass evolution were available, we found the diagram useful for the purposes of this chapter. In Figure 5.4, we have mapped the host ranges of several cicadellid species to illustrate varying degrees of host specificity.

It would clearly be useful to convert our qualitative mapping concept into a quantitative one and, as we continue to accumulate a large database of cicadellid occurrence on many grasses, we are beginning this task. A simple approach is to consider each insect occurrence as a separate event, and to weight each of the 1000-odd grass species equally. An index of specificity can then be calculated as the fraction of occurrences on the most frequently colonized host divided by all occurrences. Consideration of data for a monophagous species, *Flexamia abbreviata* [host: *Bouteloua gracilis*], yielded a value of 0.967 when computed in this way. This simple index does not weight taxonomic or phylogenetic position of the host, and unless modified, ignores geographic variation in specificity patterns (155). Weighting can be achieved by giving equal weight to each Nearctic grass tribe, recording the maximum density that a cicadellid species achieves on any member of each tribe, and computing an index from the fraction of "occupancy" of the available tribes. There are of course, many other approaches, each with its own advantages and pitfalls. The development of a suitable index (or indices) that express host breadth of the various species will enable us to use our database to test many hypotheses concerning environmental influences on cicadellid host selection.

Host Range and Evolution in Deltocephalinae

The significance of host patterns in deltocephaline leafhoppers was clearly recognized by Oman (97). Although he refrained from phylogenetic speculations, he diagrammed possible relationships among deltocephaline genera. He also pointed out the correlation between host plants and the general divisions of the deltocephaline genera, which were made on objective morphological characteristics of the male genitalia. These were (i) Y-type versus linear connective and (ii) articulation versus fusion between the connective and aedeagus. A phylogenetic sequence in which a Y-type connective is followed by a linear one and in which articulated members

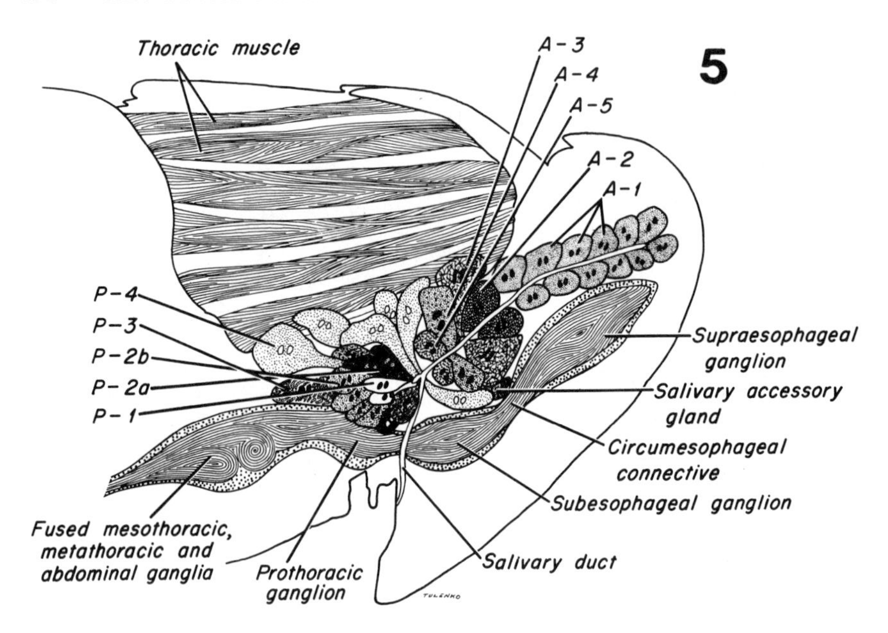

FIGURE 5.5. Salivary glands of *Colladonus montanus*. Diagrammatic representation of the ten acinus types and accessory gland.

eventually fuse appears to be most logical. Additional strength to this interpretation is provided by leafhopper salivary gland morphology. Primitive deltocephaline leafhoppers such as *Colladonus* (Figs. 5.5–5.10) and *Fieberiella* possess complex glands that consist of ten acini types in all (five types in both anterior and posterior lobes.) In contrast, more recent genera, which are reduced in size, contain much simpler glands.

A generalized view of evolution in the Deltocephalinae that follows a general trend for increasing host specialization, morphological simplification, and reduction in body size is diagrammed in Figures 5.11 and 5.12. This view assumes that primitive genera (e.g., *Fieberiella*) subsisted entirely on woody plants. A large number of genera in the primitive lineage of Deltocephalinae alternate obligately between woody and herbaceous dicots. These insects are especially prominent in savanna and ecotonal biomes. From this lineage, there appear to be several lineages that evolved in grasslands. One of these represents the hecaline leafhoppers (subfamily Hecalinae of some authors), classed on gross morphological grounds as a separate subfamily, but now seen on grounds involving both genitalic and other internal anatomical features to belong to the deltocephaline lineage. Other lineages of grass feeders in this wing include *Athysanus, Doratura, Exitianus,* and the genera of the *Stirellus* cluster. Significantly, it should be noted that the cicadellid genera in this lineage (Y-type connective articulated to the aedeagus) are diverse not only in their host relationships, but also in their geographic distribution. Over evolutionary

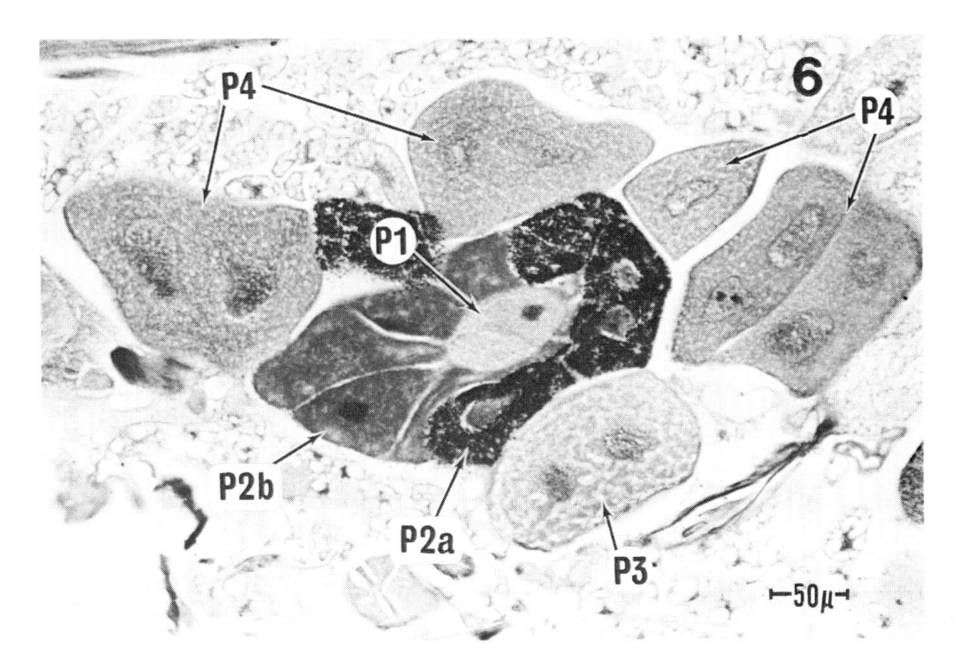

FIGURE 5.6. Posterior lobe of salivary gland of *Colladonus montanus*. Young adult fixed in Duboscq-Brasil, with Mallory's triple stain. (P-1) Cytoplasm finely granular, homogeneous with light blue stain. Compact nuclei staining red. (P-2a) Cytoplasm packed with round red-staining globules. Nuclei larger than those of P-1 or P-2b, elongated or compact, red-staining, with granular chromatin. (P-2b) Cytoplasm packed with smaller, blue-staining globules. (P-3) Cytoplasm with large aggregates containing many small vacuoles, staining blue; nuclei large. (P-4) Cytoplasm taking a light blue, yellow, or orange stain; although depicted as homogeneous and granular, often irregular. Nuclei large, with irregular outlines.

time, they have achieved intercontinental dispersal. On the other hand, groups of deltocephaline leafhoppers with linear connectives are principally Nearctic, or in some cases, Neotropical. In cases where genera are shared with Eurasia, there is evidence that dispersal occurred from the New World to the Old (120). Furthermore, these genera are entirely restricted to grasses or sedges. Thus, it is natural to speculate that these groups emerged and evolved with North American grasslands.

Evolutionary Patterns in Selected Nearctic Leafhopper Genera

THE GENUS *FLEXAMIA*

The genus *Flexamia*, more than any other, is truly representative of the North American prairie. This genus consists at present of 34 species, almost all of which are associated with at least one of the main types of

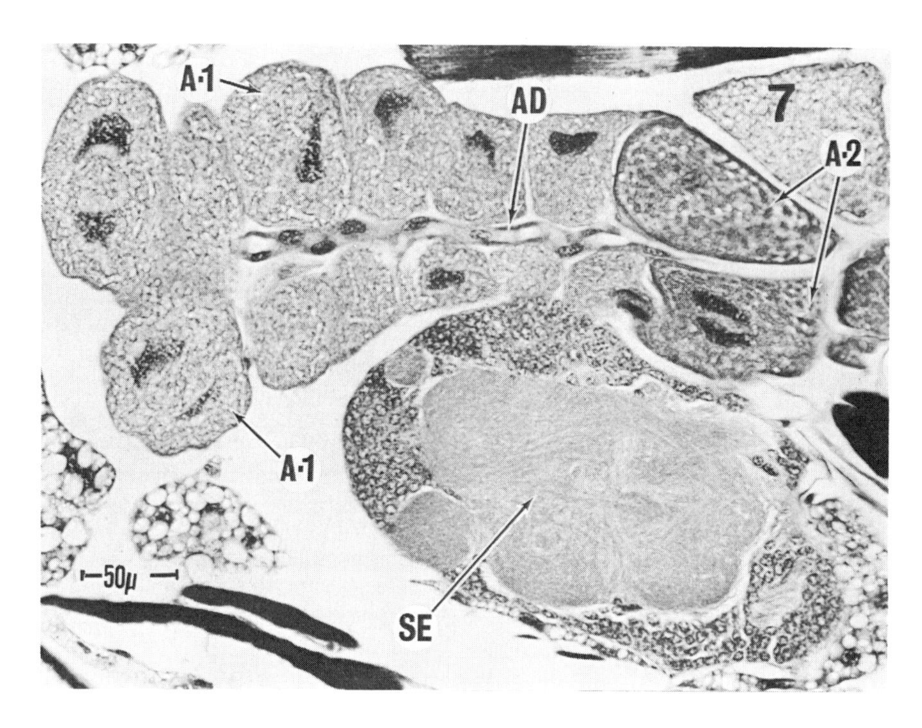

FIGURE 5.7. Salivary glands of young adult *Colladonus montanus*. Mallory's triple stain, fixed in Duboscq-Brasil. Serous cells (A-1 acini), A-2 acini, and supra-esophageal ganglion (SE). (A1) (Serous) acini. Cytoplasm taking light blue stain. Nuclei varying from round to irregular, taking reddish stain. (A-2) Cytoplasm more granular than that of A-1 cells, taking pinkish stain. Nuclei elongate and dark-staining. (AD) Anterior duct comprised of flattened cells with small red nuclei.

primordial grassland. Twenty of the species are associated with true prairie, five with the desert plains, three with southeastern grasslands, two with coastal plains of the southeast, and one with the northwestern Palouse prairie. This genus therefore probably evolved with the prairie. A brief sketch of phylogenetic trends in the genus was given by Young and Beirne in their 1959 generic revision (168). Also, H.H. Ross was especially interested in *Flexamia* and constructed a preliminary phylogeny for the genus (116; unpublished notes). We have synthesized ideas from these authors and present them, with further revisions, in Fig. 5.13. Genera *Spartopyge* (southwestern and Mexican) and *Aflexia* are thought to have diverged before the main radiation in the prairie. The genus *Alapus,* a southeastern genus, also probably diverged in the early stages of development of this lineage. At the same time, a branch involving *F. albida* and *F. serrata* (host: *Muhlenbergia*) diverged, as did *F. slossonae*. This species is at present confined to Florida, where it feeds on *Distichlis spicata*. A second lineage (*F. picta* and *F. pyrops*) also diverged during early evolution of

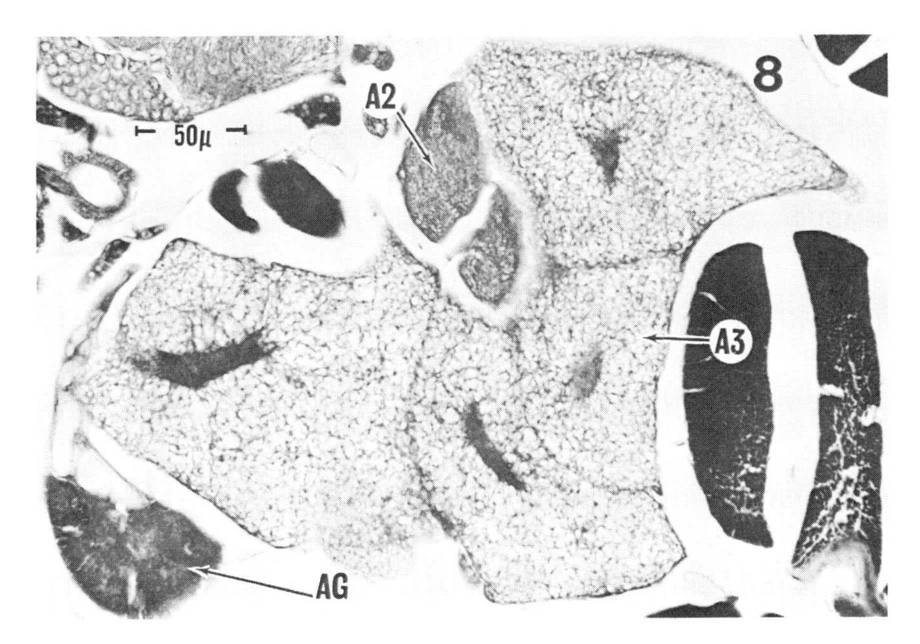

FIGURE 5.8. As for Figure 5.7, A-3 acini and accessory gland. Cytoplasm of A-3 cells (largest in gland) finely reticulate, taking light blue stain. Nuclei irregular in shape, staining dark red. Accessory gland (AG) containing eight cells, with darkly staining nuclei. Cytoplasm taking a heavy blue stain. Duct comprised of flattened cells with small red nuclei.

the genus, in association with *Aristida* spp. A group of species associated with the true prairie (i.e., *F. pectinata* and *F. abbreviata*) is associated with *Bouteloua* spp. Also, a true prairie species (*F. curvata;* host: *Buchloë dactyloides*) and a related form, *F. surcula* from subtropical Texas (host also *Buchloë*), may be relatively close to this cluster. Two forms of the desert plains, *F. doeringae* and *F. canyonensis,* also belong in this cluster. Hosts have not been identified for these species. Two species (*F. stylata* and *F. inflata*) appear to lie at the heart of the *Flexamia* lineage. These species may have somewhat wider host ranges among grasses than other *Flexamia* species; for example, *F. inflata* utilizes *Eragrostis, Muhlenbergia, Festuca,* and even *Juncus* (*Juncaceae*) as food plants. *Flexamia decora* is associated with *Muhlenbergia* spp., as is the closely related *F. imputans. Flexamia areolata,* which is closely related to *F. imputans,* specializes on *Eragrostis spectabilis.* A lineage that is closely related to *F. imputans* has given rise to a group of two species of the northern plains. The unusually patterned *F. grammica* is restricted to *Calamovilfa longifolia. Flexamia atlantica* is often found to be associated with switchgrass (*Panicum virgatum*) or big bluestem (*Andropogon gerardii*); it has dispersed as far east as New Jersey, possibly on switchgrass. Two recently

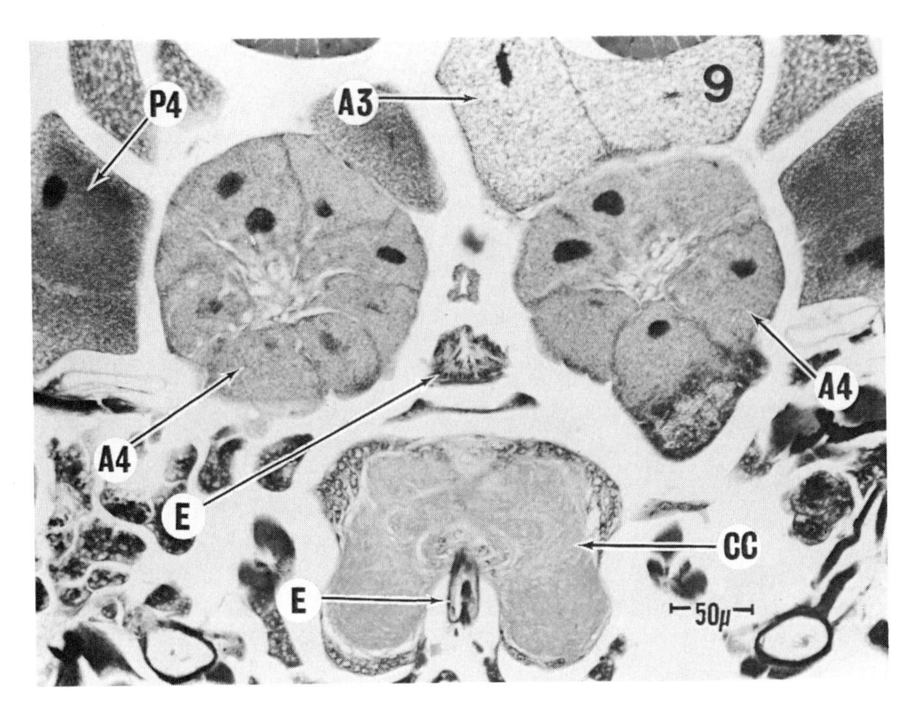

FIGURE 5.9. Acini of anterior lobe of salivary gland of young adult *Colladonus montanus*. Mallory's triple stain, fixed in Duboscq-Brasil. A-4 acini. Eight cells in a lifesaver-like configuration in a single plane. Both anterior lobes are shown. Cytoplasm taking intense blue stain. Nuclei large for the cell size, round, dark red-staining. Other structures are A-3 acini and P-4 acini, (CC) circumesophageal connective, and (E) esophagus.

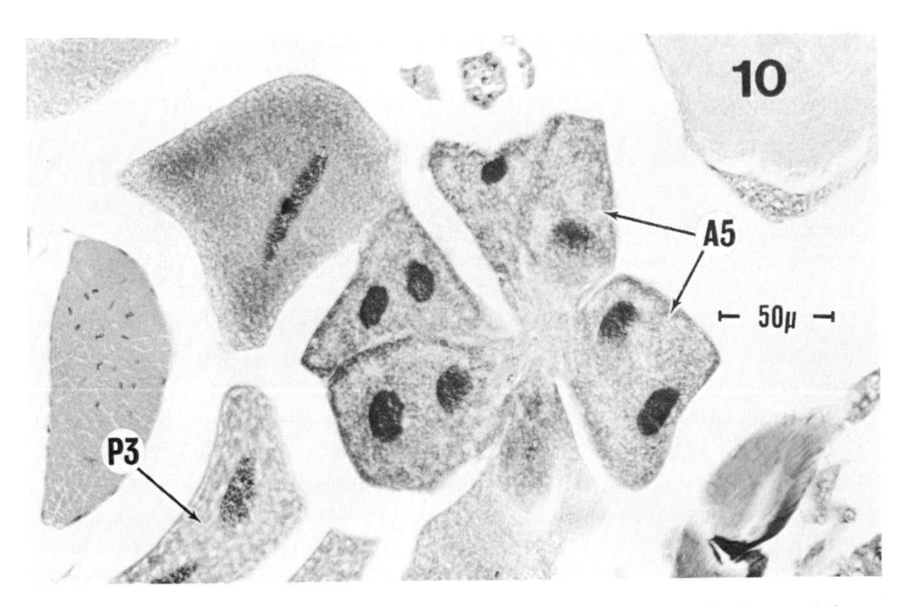

FIGURE 5.10. As in Figure 5.9, A-5 acini. Nuclei very large for cell size, staining intensely. Cytoplasm often staining orange or yellow or, more rarely, blue.

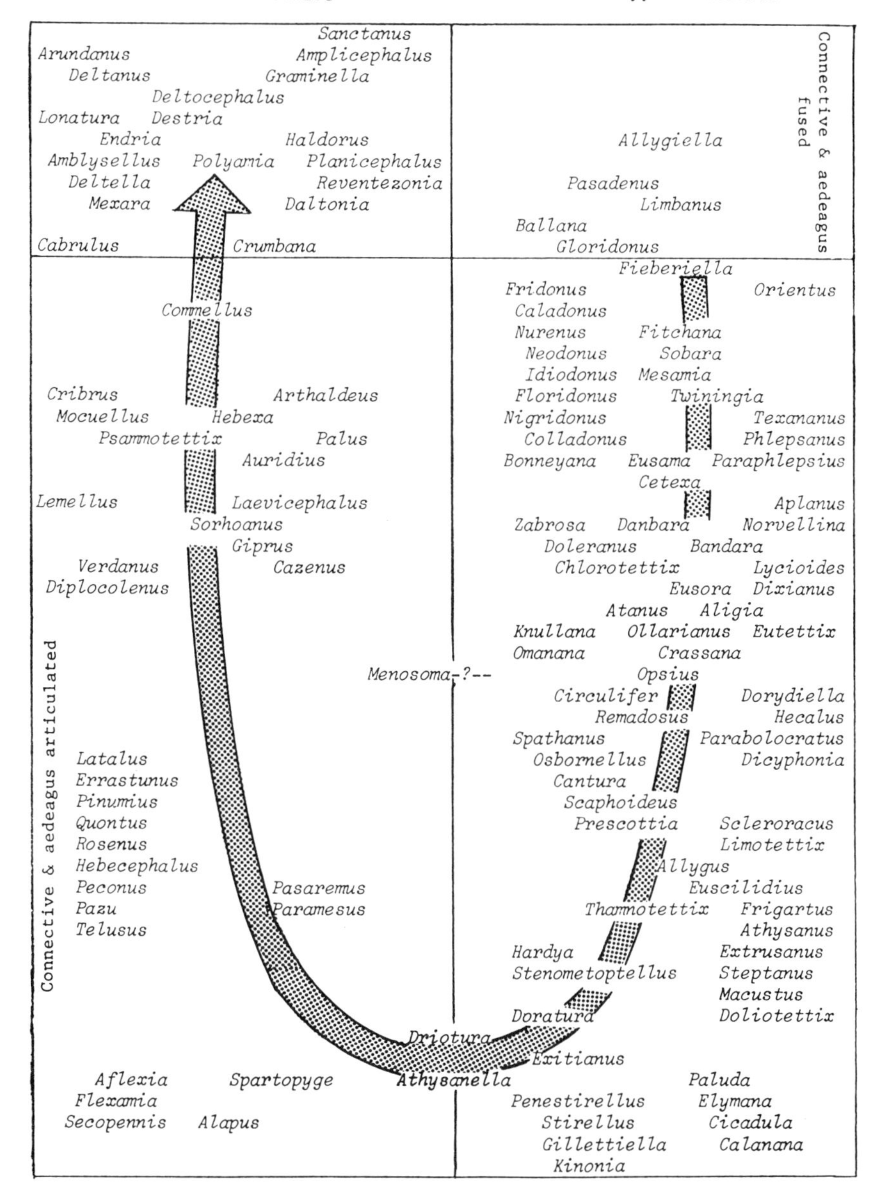

FIGURE 5.11. Two-dimensional diagram of relationships among genera of Nearctic Deltocephalini. Arrow indicates probable direction of evolution. Primitive (Y-type connective, articulated with aedeagus) genera may be worldwide in distribution. Most primitive forms are associated in some way with woody hosts; more recent genera (*Exitianus, Stirellus*) tend to be grass-feeders. Dispersal of the ancestors of these genera may have preceded continental drift. Recent (linear connective) forms are almost entirely confined to the New World, and feed on grasses (or rarely, sedges). [Modified from Oman (97).]

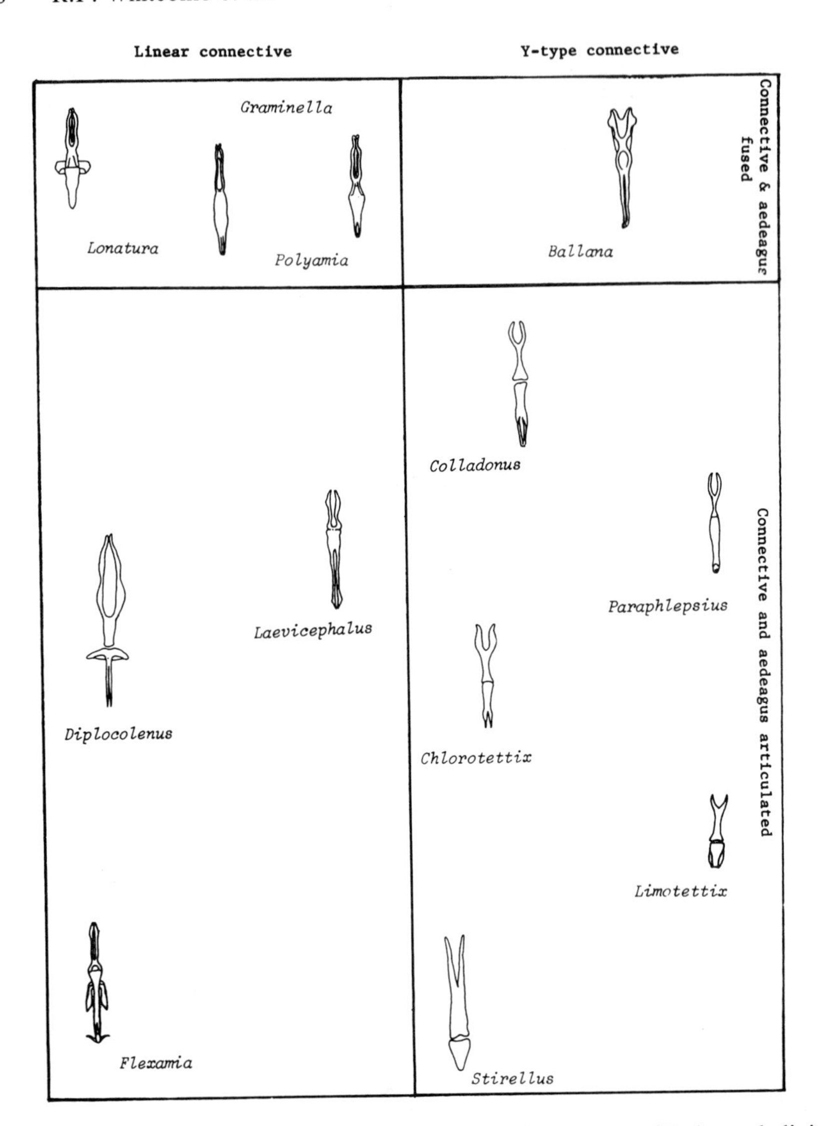

FIGURE 5.12. Male genitalic structures of representative genera of Deltocephalini. See Figure 5.11 for interpretation of connective type and their relationships to the aedeagus.

described (J.E. Lowry and H.D. Blocker, 1987) species, one (*F. arenicola*) related to *F. flexulosa* and another (*F. celata*) related to *F. stylata,* were found in our collections from sandhill muhly (*Muhlenbergia pungens*) and blowout grass (*Redfieldia flexuosa*) in the Nebraska and Colorado sand hills.

There is a large cluster of *Flexamia* species that have speciated in association with andropogonoid grasses. Throughout much of the prairie,

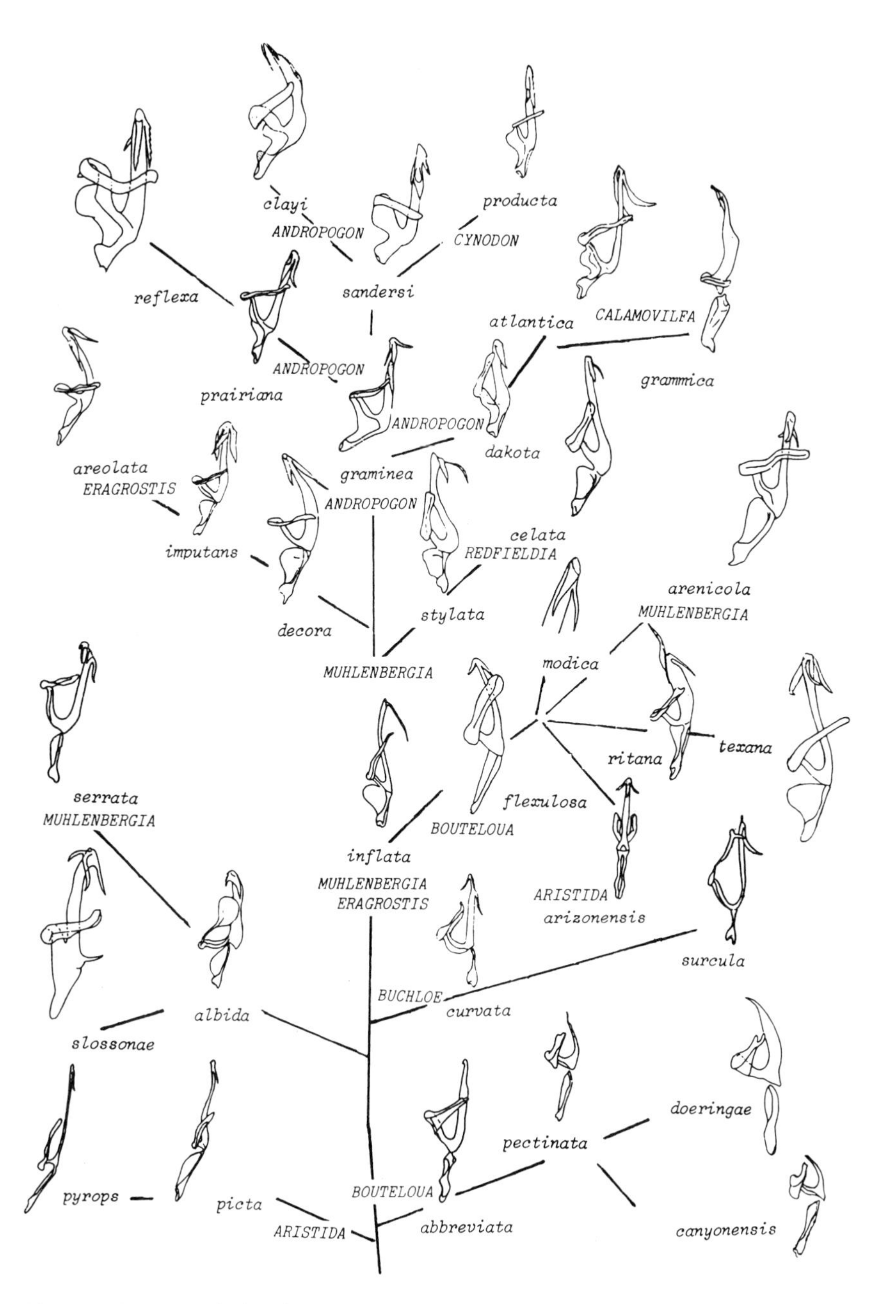

FIGURE 5.13. Evolution in the genus *Flexamia* (male genitalia). More primitive species are principally southwestern and southeastern. Other branches of the genus are northwestern, or are residents of the Nebraska sand hills region. Although a few species of the *prairiana–graminea* group are southeastern, most are associated with the true prairie. Evolutionary relationships are closely tied to food preferences. Primitive species are associated with *Bouteloua, Muhlenbergia,* and *Aristida*. Sand-hill forms are associated with *Muhlenbergia, Redfieldia,* or *Eragrostis*. Recent evolution (*prairiana–graminea* group) appears to have taken place on *Andropogon*. Distribution and host relationships are closely linked to Pleistocene events. See text for details. Artwork redrawn in part from Young and Beirne (1958).

Schizachyrium scoparium is the dominant andropogonoid grass, particularly on upland sites. In the southeast, *Andropogon virginicus* is a dominant in early fire succession and therefore probably occupied large areas of primordial habitat. In the Texas prairie, *Bothriochloa saccharoides* and others predominate, and throughout the tall-grass prairie, low areas are often occupied by *A. gerardii*. *Flexamia prairiana, reflexa, sandersi, delongi, dakota, graminea, satilla,* and *clayi* all occur on (and may be restricted to) andropogonoid grasses. Thus, the cluster of andropogonoid specialists represents the largest single species group within the genus (115). Although this group appears to have arisen relatively late in the evolution of the genus, it evolved in close association with a major group of dominant grasses (from the point of view of available biomass through recent evolutionary time) of the Nearctic prairie. Most of the species of the cluster are associated with true prairie regions, but *F. clayi* and *F. producta* are southeastern in distribution. *Flexamia clayi* was taken from *A. virginicus* in our collections from Indiana and Ohio. The primordial host or hosts for *F. producta* have not been determined, but today, this species is distributed widely on the introduced pasture grass *Cynodon dactylon* and has even been found in one instance on a stand of the introduced *Zoysia japonica*.

Flexamia Biogeography and Geological History

Distribution of *Flexamia* species today, of course, must be much different than at the time of glaciation. In fact, it is likely that the origin of certain species is postglacial. Among these might be the sand hill endemics *F. arenicola* and *F. celata,* which are associated with sand hill grasses such as *Redfieldia flexuosa* and *Muhlenbergia pungens*. Although the sand hills may never have been completely glaciated (163), they must nevertheless have been a hostile environment for grassland insect species. The present-day distribution of *Flexamia dakota* is largely that of an unglaciated portion (80) of the northern plains west and north of the sand hills. Perhaps on south-facing slopes of this region, little bluestem, the host of *F. dakota,* persisted even in the much colder climatic conditions of the Pleistocene. In general, the colder climatic conditions prevailing at that time (40) must have had profound effects on the *Flexamia* complex that inhabits andropogonoid grasses of today's landscape. It has actually been postulated, on the basis of fossil evidence, that the biota as far removed as the West Indies was drastically altered by climatic cycles of the Pleistocene (104). During these glacial maxima, for example, Iowa was basically covered by spruce forests (165); it is possible that a majority of the *Flexamia* species would have been extirpated by such a drastic climatic change. A puzzling case is presented by the relict population of the primitive *Flexamia* relative *Aflexia rubranura,* found only in the Chicago lake basin area, a region that underwent drastic changes during the glacial maxima (46). One wonders where this species or its ancestors resided during the glacial maxima.

Some of the lake basin cicadellids have counterparts in Florida or the south Atlantic coast (31). One of the most unusual *Flexamia* species, *F. slossonae,* has been found only in Florida. Again, this species is far removed from its nearest relatives (*F. albida* and *F. serrata*), which are western in distribution. Several species of *Flexamia* are endemic in southern Arizona. These may be relicts from a geological period during which the area of the southwestern United States now occupied by Sonoran and Chihuahuan desert was mesic grassland (3, 72). The reduction of total area in grasslands (115) may have had a drastic effect on the species group that now inhabits andropogonoid grasses of the prairie. One of these species, perhaps occupying a primitive position in the *Andropogon* cluster (*F. prairiana*), is now distributed from the arid regions of the Chihuahuan desert (where it occurs on *Bothriochloa* species) to the eastern tall-grass prairie. This species, which may have a somewhat wider host spectrum than other *Andropogon* specialists, appears to be the most likely of contemporary species to have survived the glaciations. Other species, such as *F. graminea* and the closely related *F. clayi* (an eastern species often found on *Andropogon virginicus*), could in fact be recently derived species that arose as the cluster recolonized vast *Andropogon* prairies left in the wake of the glacial maxima. Also, the component species of the *F. sandersi* complex (*F. sandersi, F. satilla,* and *F. delongi*) could hardly have coexisted in the small area of prairie during the maxima; rather, it is most likely that these forms, the most closely related of all the *Flexamia* species, have diverged in postglacial times. In summary, clear geological evidence, taken with the contemporary limitations of *Flexamia* habitat, suggest that Pleistocene events must have had a profound effect on the distribution of the genus, and, inevitably, on speciation and perhaps extinctions of species for which we have no record. Speciation in the genus may have occurred by both allopatric and sympatric (131, 132) mechanisms. A much better reconstruction of these events will be possible when a detailed objective phylogeny is reconstructed for the genus, and when more accurate geographic ranges and host records are available for the existing species.

The Genus *Laevicephalus*

Laevicephalus is another important cicadellid genus of the North American grasslands. The genus was reviewed by Ross and Hamilton (121), who also proposed a tentative scheme for its evolution. Basically, the suggested patterns in the group involve eight groups and/or subgroups (Fig. 5.14). The *L. sylvestris* subgroup is largely associated with the tall-grass prairie and/or savanna associations of the east and northeast (Fig. 5.15). Of the five subgroup members, host data are available for only three: *L. sylvestris* has been taken from a variety of woodland grasses; *L. acus* has also been taken from various grasses, but occurs in association with *Panicum* species and *Agrostis scabra*. In contrast to these species, which appear to be oligophagous, *L. melsheimerii* is a strict specialist on *Danthonia spicata*

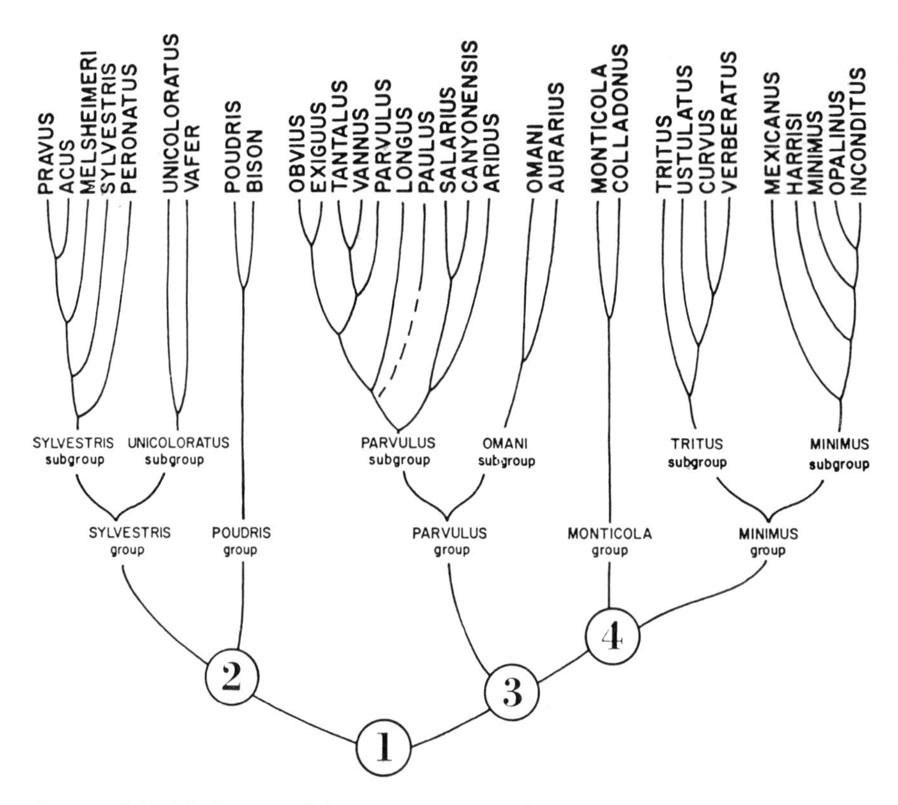

FIGURE 5.14. Phylogeny of the genus *Laevicephalus*. [From Ross and Hamilton, (121).]

and *D. compressa*. A second subgroup related to *L. sylvestris* consists of the species *L. unicoloratus* and *L. vafer*. The latter is Mexican, but *L. unicoloratus* is one of the most typical species of the tall-grass and mixed-grass prairies. It is one of the commonest members of *Andropogon* guilds, but also utilizes *Eragrostis* and certain other grasses of the tall-grass prairie. This is one of several species that commonly cross over to *Cynodon dactylon*. A second major group, consisting of two species, is the *L. poudris* group. These species, so far as is known, are true prairie species, but their hosts have not been discovered. Similarly, host data have only recently begun to be accumulated for the *L. parvulus* (short-grass and desert plains regions; *Bouteloua* and *Buchloë* are hosts for some species), *L. monticola* (southwestern and Mexican mountains), *L. tritus* (desert plains, especially Sonora), and *L. minimus* groups. The last group, although largely southwestern, did disperse to some extent into the eastern tall-grass prairie.

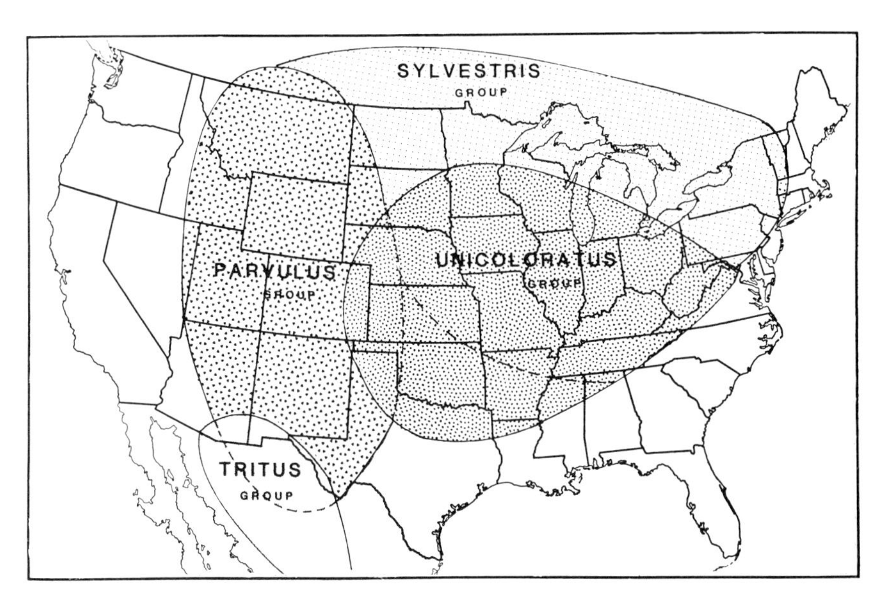

FIGURE 5.15. Geographic distribution of four species groups of *Laevicephalus*. The *sylvestris* group has various host affinities: *L. sylvestris* is associated with various pooid grasses; one species, *L. melsheimerii*, is a specialist on *Danthonia; L. unicoloratus* feeds on several grasses of the tall-grass prairie, especially *Schizachyrium scoparium* and *Sorghastrum nutans*. Members of the *parvulus* group, as far as shown, are associated with *Andropogon, Bouteloua,* and *Buchloë*. Hosts of members of the *tritus* group are unknown.

THE GENUS *POLYAMIA*

Polyamia, as currently defined, is similar in number of species to *Flexamia* and *Laevicephalus*. There are two geographic species groups. A group of seven species occurs in Arizona; although we know little about the biology of most members of this cluster, one (*P. neoyavapai*) occurs on vine mesquite (*Panicum obtusum*). The other species are largely southeastern; many of these have clearly evolved in parallel with *Dichanthelium,* a large, predominantly southeastern genus that is common on the coastal plains, in early secondary successional habitats, or in sand prairies. Some species must have been major components of savanna or ecotonal habitats. Although panicoid grasses are not well represented in the true prairie, two species (*Panicum virgatum* and *Dichanthelium scribnerianum*) in the northern prairie and *D. oligosanthes* in the southern prairie achieved important status. In the Illinois and Maryland grasslands, especially in smaller, wooded, disturbed, or sandy areas, *Polyamia* species are associated with *Dichanthelium* species, demonstrating various degrees of specificity (Table 5.6). Two exceptions to this generalization are *Polyamia weedi,* a distinct species that can be separated from other *Polyamia* species by morphological characteristics and is normally found on chloridoid rather than panicoid grasses (Fig. 5.4), and *P. caperata,* which is a regular member of cicadellid guilds of andropogonoid grasses.

The Genus *Graminella*

It is convenient in considering evolution in southeastern grasslands to consider the genus *Graminella,* which contains two of the most general of the grass feeders, *G. nigrifrons* and *G. sonorus.* These species have been taken from such a wide range of grasses in different subfamilies that it is clear that their life history strategies are significantly different from those adopted by most deltocephaline leafhoppers of the central prairie. Both of these species, for example, appear to lack the capacity for an egg diapause and therefore overwinter as adults only as far north as their cold hardiness permits. One lineage of *Graminella* appears to be specialized on *Panicum virgatum;* this includes the group comprising *G. aureovittata, G. mohri,* and *G. oquaka. Graminella villica* in northern latitudes specializes on *Paspalum* spp. (especially *P. laeve*); in the deep south, this species has apparently crossed over to *Cynodon* and probably to other grasses as well (47), but also shows an affinity to such *Paspalum* species as bahiagrass (*P. notatum*). In the French Antilles, it has been reported from pangola (11). Finally, *G. fitchii* is a specialist on *Leersia oryzoides* or *L. virginica.*

The Genus *Diplocolenus*

Ross and Hamilton (120) studied *Diplocolenus,* a small genus that consists of six species (Fig. 5.16). Three of these are associated with a variety of pooid grasses in northern prairies and forests of the Nearctic. Three other species, all of them related and therefore representing a separate lineage, are Palaearctic. This provides evidence for limited westward dispersion of deltocephaline leafhoppers with linear connectives into Eurasia (120).

The Genera *Baldulus* and *Dalbulus*

Two genera, *Baldulus* and *Dalbulus,* contain assemblages of species that include some of the most strictly specialized leafhoppers of Nearctic grasslands. They are associated with species of the Andropogoneae, largely in Mexico. In the temperate prairie, this lineage is represented only by *Baldulus tripsaci,* which occurs, with the deltocephaline *Reventezonia lawsoni,* on *Tripsacum dactyloides* in an absolutely monophagous association. These temperate deltocephaline species apparently do not occur in Mexico (L.R. Nault, personal communication), but instead are replaced

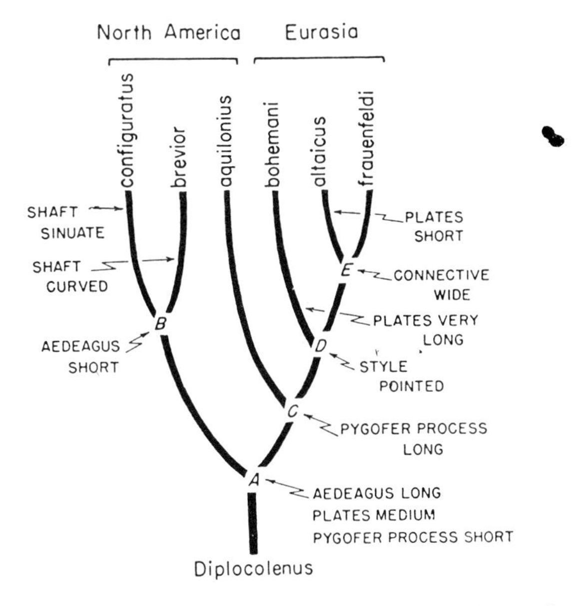

FIGURE 5.16. Phylogeny of the genus *Diplocolenus*. [Ross and Hamilton (120).]

on *Zea* and *Tripsacum* by a complex of species of the genus *Dalbulus*. Because this genus is primarily involved in transmission of the spiroplasma and mycoplasmalike organism that are the causative agents, respectively, of the destructive corn stunt and maize bushy stunt diseases (as well as certain viruses), these species have been intensively studied in recent years by Nault and his co-workers (88–93, 138).

THE GENUS *DRIOTURA*

All of the genera discussed above evolved principally or entirely with grasses. However, the story of grassland cicadellids would not be complete without discussion of genera that evolved with prairie forbs. The genus *Driotura* is a simple example, because only three species have been discovered, all of which are associated with composites. *Driotura gammaroides* is apparently a specialist on *Solidago* species, especially *S. rugosa* in the east and *S. mollis* and *S. rigida* in the central prairie. *Driotura robusta* has been found in the east on *Erigeron* (where it is uncommon) and in the plains on *Aster* spp., where it can be abundant in early succession (126). Finally, *Driotura vittata* has been taken from *Gutierrezia* [in our surveys, and in those of Lawson (79)] in the short-grass prairie regions, explaining its distribution in the more arid areas of the southwest. Some authors have reported *Driotura* species on grasses, but in our opinion such records are simply a testament to the difficulty of making collections from known hosts in mixed prairie situations.

The Genus *Paraphlepsius*

The genus *Paraphlepsius* deserves brief mention in our chapter by virtue of its prominence in savanna and ecotonal habitats (28). Many of the species of this genus appear to be associated with both woody and herbaceous vegetation. Only one species that we studied appeared to be associated entirely with grasses (*P. dentatus*, a specialist on *Leersia oryzoides*). *Paraphlepsius irroratus* is a transcontinentally distributed supertramp that occurs on grasses as well as herbaceous and woody vegetation. *Paraphlepsius altus* appears to be a habitat specialist of the tall- and mixed-grass prairies. One group of species (the *P. turpiculus* group) appears to specialize on *Carex*. The diversity of life histories and host relationships reflects the large size of the genus and the great diversity revealed by their very divergent genitalic structures (25, 54). Because the life history strategies of many *Paraphlepsius* species are very complex, they will require careful study.

Host Transfer, Ephemeral Populations, and Cicadellid Speciation

Host transfer has almost certainly been a major mechanism for cicadellid speciation. The major shifts in dominance hierarchies of plant species at a regional level (induced by climatic shifts such as those of the Pleistocene) probably afforded major opportunities for the emergence of locally isolated insect populations on new hosts. One can only presume that on an evolutionary time scale some of these populations eventually adapted to the nutritional and phenological idiosyncrasies of their new hosts. They would then, in effect, be separated from the founding population, whether or not the populations on the new and the original hosts were allopatric or sympatric. Such a case has been postulated to explain the existence of "host races" of the membracid *Enchenopa binotata*, whose biologies are closely tied to characteristics of their hosts (52, 159–162).

In the course of our collections, we have had the opportunity to observe a number of ephemeral colonizations. In Illinois, H.H. Ross observed a large population of *Flexamia inflata* on *Poa pratensis*, an unusual host for this species. Within a brief time, however, the population disappeared. At Beltsville, the Soil Conservation Service planted a drainage area in a large field to *Zoysia japonica*, as a long-term test of the efficacy of this grass for soil conservation purposes. Two *Flexamia* species—*F. inflata* and *F. producta*—became established on this grass during the period after installation. The grass was maintained without disturbances such as burning and was mowed only once annually for a decade, during which time both *Flexamia* populations persisted. Eventually, however, one of the species, *F. producta*, was extirpated.

Elymus and *Agropyron* are closely related, and hybridization is regularly observed between species of the two genera. Recently, part of *Agropyron* has been united with *Elymus* and *Elytrigia*. In Illinois, two members of the cicadellid guild of *Elymus canadensis* (*Commellus comma* and *Dorycephalus platyrhynchus*) regularly crossed over to *Agropyron repens* (153). In some years these univoltine species were able to complete their development on *A. repens* before it flowered and died. In years when maturation occurred earlier, they were not able to do so, and populations on the usual host were extirpated (or nearly so).

One way that the role of host transfer as a mediator in evolution could be "tested" would be to insert a new host suitable for insect colonization into an existing ecosystem and then observe the patterns of "crossovers," as the new dominant was colonized. This experiment has, of course, actually been done, in the introduction of such grasses as *Poa pratensis, Cynodon dactylon, Eragrostis curvula,* and *Panicum coloratum*. In each case, the consequence was colonization of the introduced grass by assemblages of cicadellids appropriate to the region.

Colonizations of bermudagrass have proved to be especially instructive. In different regions of the country *Cynodon* accumulates (Table 5.8) a representative fauna of generalists, but also certain oligophagous species (e.g., *Flexamia* spp.). Such species, of which *F. inflata* is the best example, can apparently readily establish new populations on introduced grasses. It is possible that such species may have an ideal dispersiveness and host selection behavior that would lead to speciation. *Flexamia inflata* is especially fit on *Eragrostis* spp. and *Juncus,* presumably reflecting the cooccurrence of these species in wet pockets of the prairie. In New Mexico, the host of *F. inflata* in moist habitats is *Muhlenbergia asperifolia*. In the east, this species is usually found on either *Juncus* or *Eragrostis,* without evidence of obligate alternation. It can also colonize *Festuca arundinacea, Poa,* or *Zoysia,* but populations on these unusual hosts apparently do not persist. Yet, on an evolutionary time scale, one would expect an occasional localized colony on an unusual host to become established; this would represent the first stage in speciation. It seems possible to us that species of *Flexamia* such as *F. inflata,* by their adaptation to alternative hosts among the prairie dominants, may have given rise, repeatedly, to one colonizing population after another and hence to several new species. Recent (155) extended studies of *Flexamia* biogeography promise to refine our hypotheses concerning the role of historical events of the glaciations on speciation in this interesting genus.

Conclusion

Ecology and evolution are the most basic of the agricultural sciences. In this chapter we have stressed host selection not only as a mechanism for speciation, but also as a contemporary determinant of the economic importance of cicadellid species. Many current agricultural predicaments are obviously related to major changes in habitat imposed by man on the natural landscape. Thus, biogeographical aspects of insect ecology (129) are fundamental to the study of agricultural ecology. Finally, an understanding of the ecology and evolution of the vector relationship (88, 105) is vital to application of vector research to appropriate agricultural management.

Summary

1. Evolution within many Nearctic leafhopper (Homoptera: Cicadellidae) genera paralleled the evolution of grasses and forbs in the prairie and other grassland formations.

2. Massive turnover of Nearctic grasslands from native perennial grasses to annual crop plants and introduced annual and perennial grasses induced equally massive shifts in distribution and abundance of leafhoppers.

3. Difficulties in collection of host data concerning grassland cicadellids are ascribable to (a) normal intermixing of host plants in native climax communities, (b) difficulties in identification of host plants in vegetative phase of growth, and (c) difficulties in collecting mobile insects from dense host stands.

4. Supplementation of field data with observations of insects confined on host plants in the laboratory provides important insights into the breadth of food plant tolerances of cicadellids. Usually the latter tolerances are broader than those observed in nature.

5. Although some grassland leafhoppers are monophagous and others utilize both grasses and broad-leaved dicots, most have host ranges that reflect the taxonomic position of their plant hosts.

6. Few cicadellids utilize all the diverse grasses of arundinoid, oryzoid, poocoid, chloridoid, and panicoid lines.

7. Leafhoppers of some genera have complex life cycles that involve both woody and herbaceous hosts. On either or both of the alternative hosts they may be host-specific. Since the herbaceous flora of the savanna biome was extensive, the number of potential niches in this biome is large and probably accounts for extensive speciation in *Phlepsius*-like and other genera.

8. Certain cicadellid species are habitat specialists and may regularly utilize, sequentially but in nonobligatory fashion, several dominant plants in a given habitat. Such species may be more efficient speciators than more specialized forms.

9. Evolution within the Deltocephalini was strongly influenced by host selection. Primitive genera (with Y-shaped connectives of male genitalia) inhabit woody plants or alternate between woody and herbaceous plants. Less primitive genera inhabit herbaceous dicots. Forms with linear connectives, especially those fused with the aedaegus, are almost entirely Nearctic forms that are restricted to grasses or sedges. Adaptation to grasses involved reduction of body size and simplification of salivary gland structure, including total loss of some acini types.

10. Contemporary distributions and species assemblages of cicadellids

show a strong influence of geological history. In particular, Pleistocene glaciations and associated climatic changes had dramatic effects on all elements of the prairie biota.

11. Evolution within grassland leafhopper genera was strongly influenced by availability of dominant grass species. Examples include groups of *Flexamia* such as the *picta* complex (*Aristida*), *abbreviata* complex (*Bouteloua*), *sandersi* complex (*Andropogon*), and *imputans* complex (*Muhlenbergia*). Species of *Polyamia* are associated with various *Panicum* species. Species of *Baldulus* and *Dalbulus* specialize on *Tripsacum* or *Zea*. Other genera (e.g., *Driotura*) evolved with forbs such as *Solidago* spp., *Aster* spp., and *Gutierrezia* spp.

12. Nearly all grassland leafhoppers with limited host ranges are associated with native dominant perennial plants. Assemblages on these perennials are comprised of (a) generalists that utilize a large number of plant hosts, and (b) specialists associated entirely or largely with the dominant.

13. Assemblages on introduced grasses are formed by (a) the same generalists found on a wide variety of native plants, (b) oligophagous grass-inhabiting species whose intrinsic host tolerances involve related native grasses, and (c) very rarely, introduced cicadellids that are probably associated with the grass in their region of origin.

14. Assemblages on annual grasses are formed almost entirely by generalists, whose densities are often much higher (in logarithmic phase of plant growth) than at any time on perennial hosts.

15. The cicadellid fauna of crop plants can be predicted and interpreted in terms of the outlined concepts. Understanding of the ecological and evolutionary context of pest "appearances" is a vital key to management of pest species.

Acknowledgments. We thank, first of all, H.H. Ross, who inspired earlier parts of this study, and P.W. Oman, whose support, encouragement, and discussion permitted its continuation. We thank E. Terrell for assistance with plant identifications. Helpful comments by M.F. Claridge, L.C. Hulbert, L.R. Nault, Paul Opler, Roger Peterson, Peter Price, and Nadia Waloff greatly improved the manuscript. J. Elaine Lowry performed much of the taxonomic work on *Flexamia*. Many personnel of the USDA Soil Conservation Service assisted with our field surveys.

References

1. Albertson, F.W., 1937, Ecology of mixed prairie in West Central Kansas, *Ecol. Monogr.* **7**:481–547.
2. Andrzejewska, L., 1965, Stratification and its dynamics in meadow communities of Auchenorrhyncha (Homoptera), *Ekol. Pol.* **13**:685–715.
3. Axelrod, D.L., 1950, Evolution of desert vegetation in western North America, *Contrib. Palaeontol. Carnegie Inst. Wash. Publ.* **590**:217–306.

4. Ball, E.D., 1929, A supplemental revision of the genus *Athysanus* in North America (Homoptera: Cicadellidae), *Trans. Am. Entomol. Soc.* **55:**1–81.
5. Beamer, R.H., 1936, The genus *Dicyphonia, J. Kans. Entomol. Soc.* **9:**66–71.
6. Beirne, B.B., 1952, The Nearctic species of *Macrosteles* (Homoptera: Cicadellidae), *Can. Entomol.* **84:**208–232.
7. Beirne, B.P., 1956, Leafhoppers (Homoptera: Cicadellidae) of Canada and Alaska, *Can. Entomol. (Suppl. 2)* **88:**1–180.
8. Blocker, H.D., 1967, Classification of the Western Hemisphere *Balclutha* (Homoptera: Cicadellidae), *Proc. U.S. Nat. Mus.* **122:**1–55.
9. Blocker, H.D., 1970, The genus *Stragania* in America north of Mexico (Homoptera: Cicadellidae: Iassinae), *Ann. Entomol. Soc. Am.* **63:**1424–1434.
10. Blocker, H.D., and Reed, R., 1976, Leafhopper populations of a tallgrass prairie (Homoptera: Cicadellidae): Collecting procedures and population estimates, *J. Kans. Entomol. Soc.* **49:**145–154.
11. Bonfils, J., and Delplanque, A., 1971, Distribution des principales cicadelles des prairies aux Antilles Françaises (Homoptera), *Ann. Zool. Ecol. Anim.* **3:**135–150.
12. Brown, A.L., 1950, Shrub invasion of southern Arizona desert grassland, *J. Range Manage.* **3:**172–177.
13. Bruner, W.E., 1931, The vegetation of Oklahoma, *Ecol. Monogr.* **1:**99–188.
14. Buffington, L.C., and Herbel, C.H., 1965, Vegetational changes on a semidesert grassland range from 1858 to 1963, *Ecol. Monogr.* **35:**139–164.
15. Canfield, R.H., 1948, Perennial grass composition as an indicator of condition of southwestern mixed grass ranges, *Ecology* **29:**190–204.
16. Carpenter, J.R., 1940, The grassland biome, *Ecol. Monogr.* **10:**617–684.
17. Claridge, M.F., and Reynolds, W.J., 1972, Host plant specificity, oviposition behaviour and egg parasitism of some woodland leafhoppers of the genus *Oncopsis* Hemiptera, Homoptera: Cicadellidae, *Trans. Roy. Entomol. Soc. Lond.* **124:**149–166.
18. Claridge, M.F., and Wilson, M.R., 1978, Oviposition behaviour as an ecological factor in woodland canopy leafhoppers, *Entomol. Exp. Appl.* **24:**301–309.
19. Collins, O.B., Smeins, F.E., and Riskind, D.H., 1975, Plant communities of the blackland prairie of Texas, in: M.K. Wali (ed.), *Prairie: A Multiple View,* University of North Dakota Press, Grand Forks, pp. 75–88.
20. Costello, D.F., 1944, Important species of the major forage types in Colorado and Wyoming, *Ecol. Monogr.* **14:**107–134.
21. Coupe, T.R., and Schultz, J.T., 1968a, Biology of *Endria inimica* in North Dakota, *Ann. Entomol. Soc. Am.* **61:**802–806.
22. Coupe, T.R., and Schultz J.T., 1968b, The influence of controlled environments and grass hosts on the life cycle of *Endria inimica, Ann. Entomol. Soc. Am.* **61:**74–76.
23. Coupland, R.T., 1961, A reconsideration of grassland classification in the Northern Great Plains of North America, *J. Ecol.* **49:**135–167.
24. Coupland, R.T., and Brayshaw, T.C., 1953, The fescue grassland in Saskatchewan, *Ecology* **34:**386–405.
25. Crowder, H.W., 1952, A revision of some phlepsiuslike genera of the Tribe Deltocephalini (Homoptera, Cicadellidae)in America north of Mexico, *Univ. Kans. Sci. Bull.* **35:**309–541.

26. Cwikla, P.S., and Blocker, H., 1981, An annotated list of the leafhoppers (Homoptera: Cicadellidae) from tallgrass prairie of Kansas and Oklahoma, *Trans. Kans. Acad. Sci.* **84**:89–97.
27. DeLong, D.M., 1923, The distribution of the leafhoppers of Presque Isle, Pennsylvania and their relationship to plant formations, *Ann. Entomol. Soc. Am.* **16**:363–374.
28. DeLong, D.M., 1926, Food plant and habitat notes on some North American species of *Phlepsius, Ohio J. Sci.* **26**:69–72.
29. DeLong, D.M., 1932, Some problems encountered in the estimation of insect populations by the sweeping method, *Ann. Entomol. Soc. Am.* **25**:13–17.
30. DeLong, D.M., 1941, The genus *Arundanus* in North America, *Am. Midl. Nat.* **25**:632–643.
31. DeLong, D.M., 1949, The leafhoppers or Cicadellidae of Illinois, *Ill. Nat. Hist. Surv. Bull.* **24**:93–376.
32. DeLong, D.M., 1965, Ecological aspects of North American leafhoppers and their role in agriculture, *Bull. Entomol. Soc. Am.* **11**:9–26.
33. DeLong, D.M., 1971, The bionomics of leafhoppers, *Annu. Rev. Entomol.* **16**:179–210.
34. DeLong, D.M., and Cartwright, O.L., 1926, The genus *Chlorotettix*—A study of the internal male genitalia, including the description of a new species, *Ann. Entomol. Soc. Am.* **19**:499–511.
35. Denno, R.F., 1977, Comparison of the assemblages of sap-feeding insects (Homoptera–Hemiptera) inhabiting two structurally different saltmarsh grasses in the genus *Spartina, Environ. Entomol.* **6**:359–372.
36. Denno, R.F., 1980, Ecotope differentiation in a guild of sap-feeding insects on the saltmarsh grass *Spartina patens, Ecology* **61**:702–714.
37. Denno, R.F., Raupp, M.J., and Tallamy, D.W., 1981, Organization of a guild of sap-feeding insects: Equilibrium vs. nonequilibrium coexistence, in: R.F. Denno and H. Dingle (eds.), *Insect Life History Patterns,* Springer-Verlag, New York, pp. 151–181.
38. Dethier, V.G., 1954, Evolution of feeding preferences in phytophagous insects, *Evolution* **8**:33–54.
39. Dingle, H. (ed.), 1978, *Evolution of Insect Migration and Diapause,* Springer-Verlag, New York, 284 pp.
40. Dort, W., Jr., 1970, Recurrent climatic stress on Pleistocene and recent environments, in: S.W. Dort, Jr. and J.K. Jones, Jr. (eds.), *Pleistocene and Recent Environments of the Central Great Plains,* University of Kansas Press, Lawrence, Kansas, pp. 3–7.
41. Durant, J.A., and Hepner, L.W., 1968, Occurrence of leafhoppers (Homoptera: Cicadellidae) on corn in South Carolina, *J. Ga. Entomol. Soc.* **3**:77–82.
42. Durant, J.A., 1968, Leafhopper populations on ten corn inbred lines at Florence, South Carolina, *Ann. Entomol. Soc. Am.* **61**:1433–1436.
43. Durant, J.A., 1973, Notes on factors influencing observed leafhopper (Homptera: Cicadellidae) populations densities on corn, *J. Ga. Entomol. Soc.* **8**:1–5.
44. Dyksterhuis, E.J., 1948, The vegetation of the western Cross Timbers, *Ecol. Monogr.* **18**:325–376.
45. Evans, J.W., 1963, The phylogeny of the Homoptera, *Annu. Rev. Entomol.* **8**:77–94.

46. Frye, J.C., Willman, H.B., and Black, R.F., 1965, Outline of glacial ecology of Illinois and Wisconsin, in: H.E. Wright, Jr. and D.G. Frey (eds.), *The Quaternary of the United States,* Princeton University Press, Princeton, New Jersey, pp. 43–61.
47. Genung, W.G., and Mead, F.W., 1969, Leafhopper populations (Homoptera: Cicadellidae) on five pasture grasses in the Florida everglades, *Fla. Entomol.* **52:**165–170.
48. George, J.A., and Davidson, T.R., 1959, Notes on life-history and rearing of *Colladonus clitellarius* (Say) (Homoptera: Cicadellidae), *Can. Entomol.* **91:**376–379.
49. Gilmer, R.M., Palmiter, D.H., Schaeffers, G.A., and McEwen, F.L., 1966, Insect transmission of X-disease virus of stone fruits in New York, *N.Y. Agric. Exp. Stn. Bull.* **813:**1–22.
50. Gould, F.W., and Shaw, R.B., 1983, *Grass Systematics, 2nd ed.,* Texas A&M University Press, College Station, Texas. 397 pp.
51. Greene, J.F., 1971, A revision of the Nearctic species of the genus *Psammotettix* (Homoptera: Cicadellidae), *Smithson. Contrib. Zool.* **74:**1–40.
52. Guttman, S.I., Wood, T.K., and Karlin, A., 1981, Genetic differentiation among host plant lines in the sympatric *Enchenopa binotata* Say complex (Homoptera: Membracidae), *Evolution* **35:**205–217.
53. Hamilton, K.G.A., 1975a, A review of the Northern hemisphere Aphrodina (Rhynchota: Homoptera: Cicadellidae) with special reference to the Nearctic fauna, *Can. Entomol.* **107:**1009–1027.
54. Hamilton, K.G.A., 1975a, Revision of the genera *Paraphlepsius* Baker and *Pendarus* Balb (Rhynchota: Homoptera: Cicadellidae), *Mem. Entomol. Soc. Can.* **96:**1–129.
55. Hamilton, K.G.A., 1982, Taxonomic changes in *Aphrophora* (Rhynchota: Homoptera: Cercopidae), *Can. Entomol.* **114:**1185–1189.
56. Hamilton, K.G.A., and Ross, H.H., 1972, New synonymy and descriptions of Nearctic species of the genera *Auridius* and *Hebecephalus* (Homoptera: Cicadellidae), *J. Georgia Entomol. Soc.* **7:**133–139.
57. Hamilton, K.G.A., and Ross, H.H., 1975, New species of grass-feeding deltocephaline leafhoppers with keys to the Nearctic species of *Palus* and *Rosenus* (Rhynchota: Homoptera: Cicadellidae), *Can. Entomol.* **107:**601–611.
58. Hanson, H.C., and Whitman, W., 1937, Plant succession on solenetz soils in western North Dakota, *Ecology* **18:**516–522.
59. Hanson, H.C., and Whitman, W., 1938, Characteristics of major grassland types in Western North Dakota, *Ecol. Monogr.* **8:**57–114.
60. Hardee, D.D., Forsythe, H.Y., Jr., and Gyrisco, G.G., 1963, A survey of the Hemiptera and Homoptera infesting grasses (Gramineae) in New York, *J. Econ. Entomol.* **56:**555–559.
61. Hayward, H.E., 1928, Studies of plants in the Black Hills of South Dakota, *Bot. Gaz.* **85:**353–412.
62. Henderson, I.F., and Whitaker, T.M., 1977, The efficiency of an insect suction sampler in grassland, *Ecol. Entomol.* **2:**57–60.
63. Hendrickson, G.O., 1928, Some notes on the insect fauna of an Iowa prairie, *Ann. Entomol. Soc. Am.* **21:**132–138.
64. Hendrickson, G.O., 1930, Studies on the insect fauna of Iowa prairies, *Iowa State Coll. J. Sci.* **4:**49–179.

65. Hendrickson, G.O., 1931, Further studies on the insect fauna of Iowa prairies, *Iowa State Coll. J. Sci.* **5**:195–209.
66. Hennig, W., 1981, *Insect Phylogeny,* Wiley, New York. 514 pp.
67. Hoffman, R.S., and Jones, J.K., Jr., 1970, Influence of late-glacial and post-glacial events on the distribution of recent mammals on the northern Great Plains, in: *Pleistocene and Recent Environments of the Central Great Plains,* S.W. Dort, Jr. and J.K. Jones, Jr. (eds.), University of Kansas Press, Lawrence, Kansas, pp. 355–394.
68. Hulbert, L.C., 1985, Fire effects on tallgrass prairie vegetation, in G.K. Clambey and R.H. Pemble (eds.), *The Prairie: Past, Present and Future, Proceedings* of the *Ninth North American Prairie Conference.* Tri-College University Center of Environmental Studies, Fargo, North Dakota/Moorhead, Minnesota, pp. 138–142.
69. Kelting, R.W., 1954, Effects of moderate grazing on the composition and plant production of a native tall-grass prairie in central Oklahoma, *Ecology* **35**:200–207.
70. Kendeigh, S.C., 1954, History and evaluation of various concepts of plant and animal communities in North America, *Ecology* **35**:152–171.
71. Kevan, P.G., Saville, D.B.O., and Chalener, W.G., 1975, Interrelationships of early terrestrial arthropods and plants, *Palaeontology* **18**:391–417.
72. Kottlowski, F.E., Cooley, M.E., and Ruhe, R.V., 1965, Quaternary geology of the southwest, in: H.E. Wright, Jr. and D.G. Frey, eds. *The Quaternary of the United States,* Princeton University Press, Princeton, New Jersey, pp. 287–288.
73. Kramer, J.P., 1964, New World leafhoppers of the Subfamily Agalliinae: A key to genera with records and descriptions of species (Homoptera: Cicadellidae), *Trans. Am. Entomol. Soc.* **89**:141–163.
74. Kramer, J.P., 1967a, A taxonomic study of *Graminella nigrifrons,* a vector of corn stunt disease, and its congeners in the United States (Homoptera: Cicadellidae: Deltocephalinae), *Ann. Entomol. Soc. Am.* **60**:604–616.
75. Kramer, J.P., 1967a, A taxonomic study of the brachypterous North American leafhoppers of the genus *Lonatura* (Homoptera: Cicadellidae: Deltocephalinae), *Am. Entomol. Soc. Trans.* **93**:433–462.
76. Küchler, A.W., 1964, Potential natural vegetation of the conterminous United States, Special Publication 36, American Geographical Society, Washington, D.C. 116 pp.
77. Lang, R., 1945, Density changes of native vegetation in relation to precipitation, *Bull. Univ. Wyo. Agric. Exp. Stn.* **272**:1–31.
78. Larson, F., and Whitman, W., 1942, A comparison of used and unused grassland mesas in the badlands of South Dakota, *Ecology* **23**:438–445.
79. Lawson, P.B., 1928, The genus *Driotura* Osborn and Ball and the genus *Unoka* gen. n. (Homoptera: Cicadellidae), *Ann. Entomol. Soc. Am.* **21**:449–462.
80. Lemke, R.W., Laird, W.M., Tipton, M.J., and Linvall, R.M., 1965, Quaternary geology of northern Great Plains, in: H.E. Wright, Jr., and D.G. Frey (eds.), *The Quaternary of the United States,* Princeton University Press, Princeton, New Jersey, pp. 15–27.
81. Lowry, J.E., and Blocker, H.D. 1987. Two new species of *Flexamia* from the Nebraska sand hills (Homoptera: Cicadellidae: Deltocephalinae) Proc. Entomol. Soc. Wash. **89**:57–60.

82. McClanahan, R.J., 1962, Food preferences of the six-spotted leafhopper *Macrosteles fascifrons* (Stål), *Proc. Entomol. Soc. Ontario* **93:**90–92.
83. Mengel, R.M., 1970, The North American central plains as an isolating agent in bird speciation, in: W. Dort, Jr. and J.K. Jones, Jr. (eds.), *Pleistocene and Recent Environments of the Central Great Plains,* University of Kansas Press, Lawrence, Kansas, pp. 279–340.
84. Morris, M.G., 1973, The effects of seasonal grazing on the heteroptera and auchenorrhyncha (Hemiptera) of chalk grassland, *J. Appl. Ecol.* **10:**761–780.
85. Morris, M.G., 1981, Responses of grassland invertebrates to management by cutting. III. Adverse effects on Auchenorrhyncha, *J. Appl. Ecol.* **18:**107–123.
86. Moss, E.H., and Campbell, J.A., 1947, The fescue grassland of Alberta, *Can. J. Res. C* **25:**209–227.
87. Nakamura, K., Ito, Y., Miyashita, K., and Takai, A., 1967, The estimation of population density of the green rice leafhopper, *Nephotettix cincticeps* Uhler, in spring field by the capture–recapture method, *Res. Popul. Ecol.* **9:**113–129.
88. Nault, L.R., 1983, Origins of leafhopper vectors of maize pathogens in Mesoamerica, in: *Proceedings International Maize Virus Disease Colloquium Workshop 2–6 August 1982.*
89. Nault, L.R., 1985, Evolutionary relationships between maize leafhoppers and their host plants, in: L.R. Nault and J.G. Rodriguez (eds.), *The Leafhoppers and Planthoppers,* Wiley, New York, pp. 309–330.
90. Nault, L.R., and Bradfute, O.E., 1979, Corn stunt: Involvement of a complex of leafhopper-borne pathogens, in: K. Maramorosch and K. Harris (eds.), *Leafhopper Vectors and Plant Disease Agents,* Academic Press, New York, pp. 561–586.
91. Nault, L.R., and Madden, V., 1985, Ecological strategies of *Dalbulus* leafhoppers, *Ecol. Entomol.* **10:**57–63.
92. Nault, L.R., DeLong, D.M., Triplehorn, B.W., Styer, W.E., and Doebley, J.F., 1983, More on the association of *Dalbulus* (Homoptera: Cicadellidae) with Mexican *Tripsacum* (Poaceae), including the description of two new species of leafhoppers, *Ann. Entomol. Soc. Am.* **76:**305–309.
93. Nault, L.R., Madden, L.V., Styer, W.E., Triplehorn, B.W., and Heady, S.E., 1984, Pathogenicity of corn stunt spiroplasma and maize bushy stunt mycoplasma to its leafhopper vector, *Dalbulus longulus, Phytopathology* **74:**977–979.
94. Nielson, M.W., 1962, A revision of the genus *Xerophloea* (Homoptera, Cicadellidae), *Ann. Entomol. Soc. Am.* **55:**234–244.
95. Nielson, M.W., 1965, A revision of the genus *Cuerna* (Homoptera, Cicadellidae), *U.S. Dep. Agric. Tech. Bull.* **1318:**1–48.
96. Nielson, M.W., and Toles, S.L. 1970. Interspecific hybridization in *Carneocephala, J. Kans. Entomol. Soc.* **43:**1–10.
97. Oman, P.W., 1949, The Nearctic leafhoppers. A generic classification and check list, *Mem. Wash. Entomol. Soc.* **3:**1–253.
98. Oman, P., 1970, Leafhoppers of the *Agalliopsis novella* complex, *Proc. Entomol. Soc. Wash.* **72:**1–29.
99. Osborn, H., 1912, Leafhoppers affecting cereals, grasses, and forage crops, *USDA Bur. Entomol. Bull.* **108:**1–123.

100. Osborn, H., 1928, The leafhoppers of Ohio, *Ohio Biol. Surv. Bull.* **3**:199–374.
101. Osborn, H., 1932, Leafhoppers injurious to cereal and forage crops, *USDA Circ.* **241**:1–31.
102. Osborn, H., and Ball, E.D., 1897, Studies on the life histories of grass feeding Jassidae, *Iowa Agric. Exp. Stn. Bull.* **34**:612–640.
103. Pitre, H.N., 1967, Greenhouse studies of the host range of *Dalbulus maidis,* a vector of corn stunt virus, *J. Econ. Entomol.* **60**:417–421.
104. Pregill, G.K., and Olson, S.L., 1981, Zoogeography of West Indian vertebrates in relation to Pleistocene climatic cycles, *Annu. Rev. Ecol. Syst.* **12**:75–98.
105. Purcell, A.H., 1982, Evolution of the insect vector relationship, in: M.S. Mount and G.E. Lacy (eds.), *Phytopathogenic Prokaryotes,* Vol. 1, Academic Press, New York, pp. 121–156.
106. Quinnild, C.L., and Cosby, H.E., 1958, Relicts of climax vegetation on two mesas in western North Dakota, *Ecology* **39**:29–32.
107. Raatikainen, M., 1971, Seasonal aspects of leafhopper (Hom. Auchenorrhyncha) fauna on oats, *Ann. Agric. Fenn.* **10**:1–8.
108. Raatikainen, M., and Vasarainen, A., 1976, Composition, zonation and origin of the leafhopper fauna of oat fields in Finland, *Ann. Zool. Fenn.* **13**:1–24.
109. Ramaley, F., 1939, Sand-hill vegetation of northeastern Colorado, *Ecol. Monogr.* **9**:1–51.
110. Riegel, A., 1941, Life history and habits of blue grama, *Trans. Kans. Acad. Sci.* **44**:76–83.
111. Root, R., 1967, The niche exploitation patterns of the blue-gray gnatcatcher, *Ecol. Monogr.* **37**:317–350.
112. Ross, H.H., 1958, Further comments on niches and natural coexistence, *Evolution* **12**:112–113.
113. Ross, H.H., 1965, Pleistocene events and insects, in: H.E. Wright, Jr. and D.G. Frey (eds.), *The Quaternary of the United States,* Princeton University Press, Princeton, New Jersey, pp. 583–596.
114. Ross, H.H., 1968, The evolution and dispersal of the grassland leafhoppers genus *Exitianus,* with keys to the old world species (Cicadellidae: Hemiptera), *Bull. Br. Mus. (Nat. Hist.)* **22**:1–30.
115. Ross, H.H., 1970, The ecological history of the great plains: Evidence from grassland insects, in: W. Dort Jr., and J.K. Jones, Jr. (eds.), *Pleistocene and Recent Environments of the Central Great Plains,* University of Kansas Press, Lawrence, Kansas, pp. 225–240.
116. Ross, H.H., 1971, Systematics and scientific understanding, *Bull. Entomol. Soc. Am.* **17**:15–17.
117. Ross, H.H., 1972., An uncertainty principle in evolution, *Univ. Ark. Mus. Occ. Pap.* **4**:133–164.
118. Ross, H.H., and Cooley, T.A., 1969, A new nearctic leafhopper of the genus *Flexamia* (Hemiptera: Cicadellidae), *Entomol. News* **80**:246–248.
119. Ross, H.H., and Hamilton, K.G.A., 1970, New Nearctic species of the genus *Mocuellus* with a key to Nearctic species (Hemiptera: Cicadellidae), *J. Kans. Entomol. Soc.* **43**:172–177.
120. Ross, H.H., and Hamilton, K.G.A., 1970, Phylogeny and dispersal of the

grassland leafhopper genus *Diplocolenus* (Homoptera: Cicadellidae), *Ann. Entomol. Soc. Am.* **63:**328–331.

121. Ross, H.H., and Hamilton, K.G.A., 1972, A review of the North American leafhopper genus *Laevicephalus* (Hemiptera: Cicadellidae), *Ann. Entomol. Soc. Am.* **65:**929–942.
122. Sampson, H., 1921, An ecological survey of the prairie vegetation of Illinois, *Bull. Ill. Nat. Hist. Surv.* **13:**521–577.
123. Sarvis, J.T., 1920, Composition and density of the native vegetation in the vicinity of the northern Great Plains field station, *J. Agric. Res.* **19:**63–72.
124. Schumacher, C.M., 1975, Rangeland—A proper prairie use, in: M.K. Wali (ed.), *Prairie: A Multiple View,* University of North Dakota Press, Grand Forks, North Dakota pp. 419–422.
125. Smart, J., and Hughes, N.F., 1972, The insect and the plant: Progressive palaeoelogical integration, in: H.F. van Emden (ed.), *Insect/Plant Relationships. Symposia Royal Entomological Society London 6,* Blackwell, London, pp. 143–155.
126. Smith, C.C., 1940, Biotic and physiographic succession on abandoned eroded farmland, *Ecol. Monogr.* **10:**421–484.
127. Southwood, T.R.E., 1972, The insect/plant relationship—An evolutionary perspective, in: H.F. van Emden (ed.), *Insect/Plant Relationships. Symposia Royal Entomological Society London 6,* Blackwell, London, pp. 3–30.
128. Stoddart, L.A., 1941, The palouse grassland association in northern Utah, *Ecology* **22:**158–163.
129. Strong, D.R. Jr., 1979, Biogeographic dynamics of insect–host plant communities, *Annu. Rev. Entomol.* **24:**89–119.
130. Swain, T., 1978, Plant–animal coevolution: A synoptic view of the paleozoic and mesozoic, in: J.B. Harborne (ed.), *Biochemical Aspects of Plant and Animal Coevolution,* Academic Press, New York, pp. 3–19.
131. Tauber, C.A., and Tauber, M.J., 1977a. A genetic model for sympatric speciation through habitat diversification and secondary isolation, *Nature* **268:**702–705.
132. Tauber, C.A., and Tauber, M.J., 1977b, Sympatric speciation based on allopatric changes at three loci: Evidence from natural populations in two habitats, *Science* **197:**1298–1299.
133. Tolstead, W.L., 1941, Plant communities and secondary succession in south-central South Dakota, *Ecology* **22:**322–328.
134. Tolstead, W.L., 1942, Vegetation of the northern part of Cherry County, Nebraska, *Ecol. Monogr.* **12:**255–292.
135. Tomanek, G.W., and Hulett, G.K., 1970, Effects of historical droughts on grassland vegetation in the central Great Plains, in: W. Dort, Jr. and J.K. Jones, Jr. (eds.), *Pleistocene and Recent Environments of the Central Great Plains,* University of Kansas Press, Lawrence, Kansas, pp. 203–210.
136. Tomanek, G.W., Albertson, F.W., and Riegel, A., 1955, Natural revegetation on a field abandoned for thirty-three years in central Kansas, *Ecology* **36:**407–412.
137. Transeau, E.N., 1935, The prairie peninsula, *Ecology* **16:**423–437.
138. Triplehorn, B.W., and Nault, L.R., 1985, Phylogenetic classification of the genus *Dalbulus* (Homoptera: Cicadellidae), and notes on the phylogeny of the Macrostelini, *Ann. Entomol. Soc. Am.* **78:**291–315.

139. Voigt, J.W., 1951, Vegetational changes on a 25 year subsere in the loess hill region of central Nebraska, *J. Range Manage.* **4**:254–263.
140. Voigt, J.W., and Weaver, J.E., 1951, Range condition classes of native midwestern pasture: An ecological analysis, *Ecol. Monogr.* **21**:39–60.
141. Waloff, N., 1979, Partitioning of resources by grassland leafhoppers (Auchenorrhyncha: Homoptera), *Ecol. Entomol.* **4**:134–140.
142. Waloff, N., 1980, Studies on grassland leafhoppers (Auchenorrhyncha: Homoptera) and their natural enemies, *Adv. Ecol. Res.* **11**:81–215.
143. Waloff, N., and Solomon, M.G., 1973, Leafhoppers (Auchenorrhyncha: Homoptera) of acidic grassland, *J. Appl. Ecol.* **10**:189–212.
144. Watts, J.G., 1963, Insects associated with black grama grass, *Bouteloua eriopoda, Ann. Entomol. Soc. Am.* **56**:374–379.
145. Watts, J.G., 1967, Some new and little-known insects of economic importance on range grasses, *J. Econ. Entomol.* **60**:961–963.
146. Weaver, J.E., 1942, Competition of western wheat grass with relict vegetation of prairie, *Am. J. Bot.* **29**:366–372.
147. Weaver, J.E., 1954, *North American Prairie,* Johnsen, Lincoln, Nebraska. 348 pp.
148. Weaver, J.E., 1958, Native grassland of southwestern Iowa, *Ecology* **39**:733–750.
149. Weaver, J.E., and Albertson, F.W., 1940, Deterioration of midwestern ranges, *Ecology* **21**:216–236.
150. Weaver, J.E., and Bruner, W.E., 1948, Prairies and pastures of the dissected loess plains of central Nebraska, *Ecol. Monogr.* **18**:507–549.
151. Weaver, J.E., and Bruner, W.E., 1954, Nature and place of transition from true prairie to mixed prairie, *Ecology* **35**:117–126.
152. Webb, J.J., 1941, The life history of buffalo grass, *Trans. Kans. Acad. Sci.* **44**:58–71.
153. Whitcomb, R.F., 1957, Host relationships of some grass-inhabiting leafhoppers, M.S. thesis, University of Illinois. 37 pp.
154. Whitcomb, R.F., 1972, Transmission of viruses and mycoplasma by the Auchenorrhyncha: Homoptera, in: C.I. Kado and H.O. Agrawal (eds.), *Principles in Plant Virology,* Reinhold, New York, pp. 168–203.
155. Whitcomb, R.F., Hicks, A.L., Lynn, D.E., Allred, K., and Blocker, H.D., 1987, Geographic variation in host relationships of leafhoppers (Homoptera: Cicadellidae) in North American grasslands in: M. Wilson and L.R. Nault (eds.), *Second International Workshop on Leafhoppers and Planthoppers of Economic Importance,* Commonwealth Institute of Entomology, London (in press).
156. Whitfield, C.J., and Anderson, H.L., 1938, Secondary succession in the desert plains grassland, *Ecology* **19**:171–180.
157. Whitfield, C.J., and Beutner, E.L., 1938, Natural vegetation in the desert plains grassland, *Ecology* **19**:26–37.
158. Whitford, P.B., 1958, A study of prairie remnants in southeastern Wisconsin, *Ecology* **39**:727–733.
159. Wood, T.K., 1980, Divergence in the *Enchenopa binotata* Say complex (Homoptera: Membracidae) effected by host plant adaptation, *Evolution* **34**:147–160.
160. Wood, T.K., and Guttman, S.I., 1981a, The role of host plants in the spe-

ciation of treehoppers: An example from the *Enchenopa binotata* complex, in: R.F. Denno and H. Dingle (eds.), *Insect Life History Patterns,* Springer-Verlag, New York, pp. 39–54.

161. Wood, T.K., and Guttman, S.I., 1981b, Ecological and behavioral basis for reproductive isolation in the sympatric *Enchenopa binotata* complex (Homoptera: Membracidae), *Evolution* **36:**233–242.
162. Wood, T.K., and Guttman, S.I., 1983, *Enchenopa binotata* complex: Sympatric speciation?, *Science* **220:**310–312.
163. Wright, H.E., Jr., 1970, Vegetational history of the central plains, in: W. Dort, Jr. and J.K. Jones, Jr. (eds.), *Pleistocene and Recent Environments of the Central Great Plains,* University of Kansas Press, Lawrence, Kansas, pp. 157–172.
164. Wright, J.C., and Wright, E.A., 1948, Grassland types of south central Montana, *Ecology* **29:**449–460.
165. Wright, H.E., Jr., and Ruhe, R.V., 1965, Glaciation of Minnesota and Iowa. in: H.E. Wright, Jr. and D.G. Frey (eds.), *The Quaternary of the United States,* Princeton University Press, Princeton, New Jersey, pp. 29–41.
166. Young, D.A., 1959, A review of leafhoppers in the genus *Draeculacephala, U.S. Dep. Agric. Tech. Bull.* **1198:**1–32.
167. Young, D.A., 1977, Taxonomic study of the Cicadellinae (Homoptera: Cicadellidae) Part 2. New World Cicadellini and the genus *Cicadella, N. Car. Agric. Exp. Stn. Tech. Bull.* **239:**1–1135.
168. Young, D.A., Jr., and Beirne, B.P., 1958, A taxonomic revision of the leafhopper genus *Flexamia* and a new related genus, *U.S. Dep. Agric. Tech. Bull.* **1173:**1–53.

6
Feeding Strategies and the Guild Concept Among Vascular Feeding Insects and Microorganisms

David W. Tonkyn and Robert F. Whitcomb

Introduction

Two large orders of insects—the Hemiptera and Homoptera—possess mouthparts that are highly modified for piercing tissues and extracting their fluid contents. Food sought by hemipterans may be the blood of vertebrates [e.g., Cimicidae (bed bugs)], the hemolymph of other invertebrates [e.g., Reduviidae (ambush bugs) and Gerridae (water striders)] or the cytoplasm and sap of plants [many Miridae (plant bugs) and Pentatomidae (stink bugs) (76)]. The Homoptera are strictly plant feeders (23), the only exceptions being several species of aphids with predatory early instars (2).

Cobben (23) studied the morphology of the piercing–sucking mouthparts in relation to the evolutionary history of the two orders. Carter (17), Miles (74), Forbes (40), and Forbes and MacCarthy (41) have also reviewed this subject. Despite the great diversity of tissues from which food is derived, the piercing–sucking mouthparts of these orders have the same general structure. The needlelike stylet bundle that actually penetrates the tissue consists of two pairs of stylets. The inner pair or maxillary stylets interlock in such a way that two channels are formed between them. The smaller channel is used to inject saliva into the food; the larger one is the food canal through which liquid food is ingested. The inner, maxillary stylets are enclosed and supported by the outer pair of mandibular stylets. The latter are typically barbed on their ends, to anchor the stylet bundle in the prey tissue or, in some taxa, to rip and tear cells apart so their contents can be ingested.

Although the actual feeding sites of the insects in plant tissues are not

---

David W. Tonkyn, Department of Biological Sciences, Clemson University, Clemson, South Carolina 29634, USA.
Robert F. Whitcomb, Insect Pathology Laboratory, Agricultural Research Service, U.S. Department of Agriculture, Beltsville, Maryland 20705, USA.

visible, it is straightforward to kill the insects while they are feeding, and subsequently to section the plant tissue to determine the location of the stylets. In addition, all Homoptera and some Hemiptera [Pentamorpha (stinkbugs)] secrete a saliva that gels rapidly as their stylets penetrate the plant (74). The gelled saliva forms a sheath about the stylets and remains in place after the insect has finished feeding. This so-called stylet sheath is readily stained in plant sections, and provides, for most species, a second, unambiguous view of the actual tissue fed upon and the route through the plant taken to reach it. The exceptions are aphids and perhaps some mesophyll feeders that extend their stylets beyond the sheath during feeding (61, 75; D. Ullman, personal communication).

Studies of the stylets and stylet sheaths in plant tissues of Homoptera and phytophagous Hemiptera have shown that there are three distinct feeding sites of piercing–sucking insects. They are (1) the xylem and (2) phloem vascular tissues, and (3) the nonvascular tissues, including the mesophyll of leaves and the endosperm of seeds. Each presents unique problems in terms of tissue location and food quality and quantity, and requires adapted feeding modes. We now wish to review these three feeding modes and to discover why the tissue selected tends to be a conservative character in evolution, often characteristic of entire families or subfamilies. We shall also review exceptions to these rules: species that are tissue generalists, and specialists that have switched to feed on tissues that are atypical of their taxa. Finally, we shall discuss the pathogens that also inhabit these plant tissues. These microorganisms also range from generalists to specialists on xylem or phloem, though they can subdivide these tissues on a much finer spatial and temporal scale than can the insects. Therefore, insect and microorganism specialists on the same plant tissues are best considered as occupying different, only partially overlapping guilds.

Xylem Feeders

The xylem tissue carries nutrients and water in a continuous stream from the root to the aerial parts of the plant. Since xylem sap is easily collected in large quantities, as guttation liquids or as exudate from wounds (e.g., sugar maple sap) or cut stems, its chemistry is well known. Although it contains a wide variety of organic and inorganic nutrients, they occur in very low concentrations and in proportions unlike those in phloem or parenchyma (see 11, 19, 87, 88, 100, 102 for reviews).

For example, nitrogen is absorbed by the roots as nitrate or occasionally ammonia. It is partially or completely converted into amino acids and amides before being transported in the xylem. Typically only seven or eight amino acids are detectable in any sample, and the bulk of the nitrogen, up to 95%, may be bound in aspartic or glutamic acids and their amides.

Not only is this an unbalanced source of amino acids for insects (24), but it is also a very poor one. Xylem sap usually contains only 0.01–0.15% nitrogen (w/v), and may have as little as 0.0002% (w/v) (57). Amino acids may nevertheless constitute 98% of all the organic material in the xylem sap (133).

Sucrose and other sugars are often, but not always, detectable in xylem sap; reported concentrations vary from 0% (w/v) in broccoli (*Brassica oleracea*) (87), to 0.0005% (w/v) in the bean *Vicia faba* (57) and 0.5% (w/v) in the sugar beet (*Beta vulgaris*) (37). The various minerals absorbed by the roots, such as calcium, iron, phosphorus, potassium, and sulfur, are also carried upward in the xylem. Alkaloids and other secondary compounds may also be present (100).

There is little nutritional requirement for xylem-feeding insects to be host-specific. Since xylem cells are large, xylem feeders can have relatively large stylets, which can be used to probe rapidly for food. Xylem sap is chemically quite simple, with most organic constituents already in the form in which they can be absorbed across the gut wall. The sap does not vary greatly between plants, except when it carries secondary defensive chemicals. Therefore, xylem feeders should have little difficulty in locating and digesting their food, regardless of which plant species it is in. Of course, once a xylem cell is located, the insect can feed for hours or even days without moving.

Although xylem feeders need not be host-specific, there is good reason to expect them to be tissue-specific. After all, there are a number of barriers to feeding on xylem sap, in addition to the initial one of being able to insert mouthparts nondestructively into single xylem cells. The next barrier to xylem feeding derives from the fact that xylem sap is under tension, and must be pumped from the plants. As a result, xylem-feeding insects are characterized by an enlarged clypeal region that supports the large cibarial muscles required to suck xylem sap (84). Raven (100; see also 64) analyzed the energetics of this process in detail; he showed that the cibarial pump consumes a significant fraction of the energy that the insects derive from the xylem sap. Furthermore, this cost is greater at higher feeding sites on plants, due to increased xylem tension and decreased nutrient levels. This, Raven suggested, may explain why the nymphs of many Cercopidae and all Cicadidae feed on xylem in the roots.

The extremely low nutrient concentrations in xylem sap present another obstacle to feeding. Although there is less nitrogen in xylem sap than in any other plant tissue except wood, carbohydrates, not nitrogen, are probably the limiting compounds for xylem sap feeders (66, 100). Xylem-feeding insects must pump and ingest extraordinary amounts of this fluid to obtain sufficient food energy. In addition, these insects have specialized guts called filter chambers that can extract nutrients from this dilute sap against the osmotic potential without simultaneously absorbing lethal amounts of water (19, 45 for reviews). This is a critical factor, since the volume of sap ingested far exceeds the volume of the insects. Adult spittlebugs [*Philaenus spumarius* (L.)] excrete up to 2.1 g of processed xylem sap, 280 times their own fresh body weight, per day (57). The leafhopper *Draeculacephala minerva* Ball processes 2.5 g of fluid per day (58), while another leafhopper, *Graphocephala atropunctata* (Signoret) (= *Hordnia circellata*) ingests 100–1000 times its own fresh body weight per day (77, 79). These feeding rates are among the highest in the animal kingdom.

These two predictions—of host generality but tissue specificity—among xylem feeders are generally borne out in nature. The xylem feeders include most of the spittlebugs or froghoppers (Homoptera: Cercopoidea) (12, 32, 48, 56, 57, 83, 123, 133), all cicadas (Homoptera: Cicadidae) (19, 131), and all leafhoppers (Homoptera: Cicadellidae) in the tribe Cicadellinae and probably those in the related tribes Evacanthinae and Mileewaninae as well (15, 38, 58, 84, 116, 123). The spittlebugs, in particular, are famous for their catholic feeding habits. The meadow spittlebug *P. spumarius* = *leucopthalmus* was reported feeding on 91 species of plants in 29 families at one site in Ontario, including "all dicotyledonous herbs of sufficient size" except milkweed (*Asclepias syriaca* L.), as well as woody plants, a sedge, and even an equisetum (98). This same insect feeds on at least 165 host plants in Finland (49), although other spittlebugs are even more polyphagous at local sites (49).

Phloem Feeders

The phloem tissue provides an alternative, very different resource for piercing–sucking insects. Efficient use of this resource requires specialized morphological, physiological, and behavioral adaptations that differ sharply from those required for xylem feeding. The phloem feeders are the best known of the three guilds discussed here, and the nature of their food supply is critical to their ecology.

Phloem sap is more difficult to acquire in pure form than is xylem sap. Its chemistry has been studied in detail for a few plants, particularly in relation to carbohydrates (124) and nitrogenous compounds (65, 87, 88, 137). Sucrose and other oligosaccharides, the photosynthate from the leaves, constitute 90% or more of the organic material in the phloem sap. These sugars are usually found in concentrations of 10–25% (w/v), roughly three orders of magnitude greater than their concentrations in the xylem sap. Nitrogen is available in the phloem in the form of amino acids and their amides, as in xylem sap. However, a greater variety of amino acids occur in the phloem, and the total amino acid concentration is slightly higher than that in the xylem, reaching 0.03–0.4% (w/v). Many other compounds, including organic acids and phosphates, sterols, vitamins, and even large amounts of ATP are also present. The phloem is more than just a transport system for the plant's energy molecules; recent work suggests that it is also a medium of active biosynthesis of particular compounds (102).

The harvesting of phloem sap presents a number of unique problems to insects, different from those facing xylem feeders. First, the sieve tube cells through which sap travels are tiny; if an insect is to tap this vascular stream without destroying it, the insect's stylets must also be tiny. Such delicate stylets cannot be plunged randomly through plant tissue in search of the phloem. It is common for phloem feeders, especially smaller ones such as aphids and coccids, to insert the stylets between cells, where they follow an extracellular path to the food. This penetration is typically aided by enzymes contained in the saliva. It may take only a few minutes, but often requires several hours to reach the phloem. There is even one report of aphids requiring 3 days to penetrate to the phloem tissue of their woody host (92). The insects can withdraw their stylets rapidly when disturbed, but this can be expensive in terms of lost feeding time (92). Phloem feeding favors a sedentary habit (see 114 for discussion); it is fortunate that phloem is so rich in sugars, which, when excreted, can be used to "buy" the protection of ants (125).

A second problem for phloem-feeding insects is that of locating the small bundles of sieve tubes within the plants. Indeed, when phloem feeders are placed on unsuitable hosts, the insects often do not locate or ingest from the phloem, despite repeated probing (13, 70, 92 for review). In one case, the stylets were observed to pass within 3–5 μm of the phloem, without recognition (72). It is clear that in suitable host plants, the stylets move nonrandomly; there are frequent reports of the stylets turning directly toward the phloem while within the plant, or of being withdrawn and inserted repeatedly about a vascular bundle until a sieve tube is located (92). This nonrandom probing is probably mediated by both physical and chemical stimuli within the plants. The stylets of aphids (86, 127), leafhoppers (42), planthoppers (43), and psyllids and whiteflies (39) are innervated, apparently with proprioceptors that direct stylet movements within plants (9, 89, 127). In addition, aphids (71, 128), leafhoppers (4, 5), and psyllids (119) have precibarial sensilla that probably mediate chemical cues for tissue location (6). These features are not unique to phloem-feeding Homoptera; however, phloem feeders may be more dependent on internal plant cues to locate their food than are mesophyll and perhaps xylem feeders.

Unlike xylem feeders, phloem feeders need not actively ingest their food. Phloem sap rises in stems at rates of 1 m or more per hour, under pressures of 20–40 atm. When aphid stylets are cut off, phloem sap continues to flow out the cut stylet tip, at the same rate at which intact aphids excrete sap, and at rates comparable to the maximum observed for phloem sap within plants (77). Once the insects have inserted their stylets into the phloem, the sap is forced into their guts. The insects can regulate this flow, using their precibarial valve (4, 5, 71, 119), but given their poorly developed cibarial pumps, it is not clear that they ever need to augment the flow (92, 100).

If insects required only carbohydrates in their diets (as is often assumed in animal foraging theory), a phloem sap diet of 10–25% (w/v) sugar would be superb. In fact, phloem-feeding insects consume far more sugar than they use, and excrete as much as 95% of it as honeydew (77, 78). Feeding rates vary from 0.57 mg/day for the leafhopper *Orosius argentatus* (Evans) (27) to 1.6 mg/day for the aphid *Myzus persicae* (Sulz.) (28) and 48 mg/day for the large willow aphid *Tuberolachnus salignus* (Gmelin) (77; see 57, 92 for reviews). These rates represent between 2 and 32 times the fresh body weights of the insects per day, which are lower than those recorded for xylem feeders, but are still high. There is little doubt that phloem-feeding insects must maintain such high feeding rates, and waste so much carbohydrate, in order to obtain sufficient amounts of essential nitrogen (see reviews 66, 73, 120, 121, 132).

The high sugar content of phloem sap gives it a high osmotic potential; ingested sap will tend to absorb water across the gut wall and could dehydrate the insects. This problem is magnified greatly by the large volumes of food ingested. Not surprisingly, the digestive tracts of phloem-feeding insects are specialized to handle both the large volumes and their high osmolalities (45). Aphids convert as much as 40% of the sucrose into the trisaccharide melezitose, before then excreting it. Owen (85) interpreted this as a highly sophisticated form of mutualism, in which the aphids fed nitrogen-fixing bacteria in the soil below, ultimately benefitting the plant on which they or their progeny were feeding. An alternative hypothesis with some experimental justification is that the aphids convert sucrose to melezitose to reduce the osmotic pressure across their gut wall.

Phloem feeding is the predominant pattern throughout the suborder Sternorrhyncha. This includes psyllids (34, 35, 53, 126, 136), whiteflies (50, 52, 90, 107; see also 3 and 14, both cited in 122), most aphids (92 for review), and many scale insects (10, 31, 44, 107). Phloem feeding is also characteristic of certain groups in the other major suborder, the Auchenorrhyncha. These are the treehoppers (Membracidae) (60), the planthoppers (Fulgoroidea) (17, 38, 108), and most leafhoppers (8, 15, 29, 36, 38, 60, 97, 103, 111, 123). Phytophagous hemipterans do not typically feed in the phloem, although they may ingest small amounts of phloem sap.

Unlike xylem feeders, phloem-feeding insects tend to have narrow host ranges. For example, only one psyllid species in the world is known to feed on plants of two families, while 76 of the 80 British species and 86 of the 110 Czechoslovakian species are restricted to feeding on plants of a single genus (30, 54). Also, of the 445 European species of aphids that lack an obligate alternation of hosts, 407 and 440, respectively, feed on plants of a single genus and family (30). This extreme host specificity may result from either of two factors discussed above. First, the internal plant cues employed by the insects to locate and recognize the phloem apparently differ among plant species, preventing efficient foraging on more than a few, similar host species (13, 70, 82, 115). Also, the insects may be under strong selection to synchronize their life histories with that of particular host plants in order to exploit brief peaks of nitrogen availability in the phloem of those plants (73, 93, 94, 113).

Mesophyll Feeders

Insects that feed on the mesophyll tissues of leaves require a different set of adaptations than do vascular feeders. There is no difficulty in locating the tissue, since, apart from the veins, it consists essentially of the whole leaf. Nor is the tissue inaccessible; even first-instar nymphs of the leafhopper *Eupteryx melissae* Curtis have stylets longer than the average thickness of their host plant's leaves (91). In comparison with xylem or phloem sap, the quality of the food ingested is very high. The cytoplasm, with its content of nucleoplasm and organelles, is richly endowed with a balanced mixture of proteins, carbohydrates, lipids, nucleic acids, vitamins, and minerals. The challenge for mesophyll feeders is to obtain enough of this rich food. This problem has been overcome in different ways by homopteran and hemipteran mesophyll feeders.

Many homopteran mesophyll feeders ingest the contents of only one cell at a time. Pollard (91) has provided an excellent review of the morphological and behavioral adaptations that allow *E. melissae* to obtain sufficient food with this strategy. Not surprisingly, these adaptations allow for the rapid location and ingestion of the contents of individual cells. These leafhoppers have relatively large stylet bundles that are plunged rapidly into the leaf. The stylets do not insinuate themselves slowly between cells using digestive enzymes, as with many phloem feeders, but simply pierce intervening cells on their way to the food. The maxillary stylets first puncture and then rapidly empty the cell; a third-instar nymph can empty a typical palisade cell ($5 \times 10^{-6} mm^3$) in about 10 sec. Chloroplasts are larger than the food canal, 5.1 by 3.4 μm versus 3.6 by 2.0 μm, and are apparently disrupted before being ingested. Once a cell has been emptied, the stylets can be rotated and maneuvered to penetrate successively a large number of nearby cells, without being removed from the tissue. Air rushes into the emptied cells, leaving a characteristic clear spot or stippling of the leaf. Individual adults of *E. melissae* feed for about 7 min before removing their stylets and moving to another site on the leaf.

In the Homoptera, mesophyll feeding is characteristic of but not restricted to the leafhopper subfamily Typhlocybinae (15, 29, 46, 47, 55, 91, 97, 103, 106, 107, 111, 123). This is an advanced subfamily of leafhoppers on morphological grounds, suggesting that mesophyll feeding is not the ancestral pattern (but see 100). Mesophyll feeding is also characteristic of some scale insects (Coccoidea) (16, 31, 105, 107) and of many root-feeding aphids of the family Adelgidae (92). These Homoptera are all small, perhaps representing one constraint of feeding on mesophyll cells one at a time. Mesophyll feeders have narrow to moderate host ranges (20, 21, 67, 110). Of all the Homoptera, these insects are most likely to contact and ingest plant secondary compounds, a circumstance that may play a major role in restricting host ranges.

The phytophagous Hemiptera have a different solution to the problem of obtaining adequate amounts of nutrients from mesophyll. Mirid (plant-) bugs, Tingid (lace-) bugs, and Pentamorpha (stinkbugs and allies) all have what Cobben (23) calls a "predatory" feeding habit. These Hemiptera use their barbed stylets to slash back and forth, lacerating large numbers of cells at a time. The contents of these cells are then flushed out with saliva and ingested. This is a far less discriminating feeding mode, but it permits higher rates of food intake and is compatible with the body sizes in excess of those of homopteran mesophyll feeders. The large stylets that slice through plant tissue so easily are also capable of piercing and feeding in animal tissues. In fact, of all the herbivorous insects with piercing–sucking mouthparts, only these large mesophyll-feeding hemipterans are commonly facultative predators of other insects.

Overlap Between Feeding Modes

Because the types of feeding we have discussed here are so different, the insect strategists that utilize them can be considered to be members of three distinct feeding guilds. For example, mesophyll feeders may lack the behavioral, morphological, and physiological adaptations for gentle tapping of vascular tissue or for ingesting and processing large amounts of nutritionally poor food. Not surprisingly, when five species of mesophyll-feeding leafhoppers were confined to the stems, petioles, or leaf midveins of their host plants, they fed for only a few hours, and usually died within a day (106). Of course, mesophyll feeders might ingest the contents of vascular cells if their stylets encountered them by chance. This is more likely to occur among the mesophyll-feeding hemipterans, whose stylets slice into large areas of tissue at each feeding, than among the leafhoppers that usually feed on single cells in the "islets" between minor veins (91). Even in that case, however, such incidental feeding (or perhaps destruction of) occasional vascular cells leaves the vascular system essentially intact.

Similarly, many vascular feeders do not feed to any significant degree in mesophyll tissues. For one thing, they are unlikely to be able to ingest mesophyll sap as rapidly as can mesophyll specialists. In addition, they frequently lack the nutritional capability to utilize mesophyll cell sap food completely if it were ingested. For example, xylem and phloem feeders have only a few enzymes for hydrolyzing sugars in their saliva; they lack the esterases, proteinases, amylases, phosphatases, and phosphorylases that mesophyll feeders possess (74, 103). These indications are supported by experimental results. Phloem-feeding leafhoppers died more rapidly when fed an artificial diet of starch, gluten, and oil emulsion than when fed on simple sugars and tryptophane (103). When confined to mesophyll, the potato leafhopper [*Empoasca fabae* (Harris)] died almost as fast as if no food were offered; it normally feeds on phloem sap (106). In a more detailed study with yet another phloem-feeding leafhopper [*Circulifer tenellus* (Baker)], Bennet (8) showed that when caged on mesophyll, the insects died slightly faster than when fed on pure water, and not quite as fast as when starved completely. Survival on the phloem tissue was quite high. No such experiments have been performed with xylem feeders, but they would likely show the same result.

There is often little overlap between the vascular and nonvascular feeding Homoptera and phytophagous Hemiptera. In addition, the vascular feeders themselves consist of two groups with little in common. Phloem feeders passively ingest a nutritionally rich sap at the rate of 2–32 times their body weight per day. Xylem feeders, with the assistance of powerful muscles, suck from 100 to 1000 times their body weight per day of a nutritionally dilute sap. The guts and gut endosymbionts of the insects differ, and are adapted for their special foods (45). As a result, artificial diets for the two kinds of insects are vastly different (116). In addition, they face opposite water balance problems from the ingestion of such large amounts of fluid. Phloem feeders must have special adaptations to avoid dehydration, while xylem feeders must avoid absorbing too much water.

There is direct evidence that phloem and xylem feeders often do not overlap in diet. Examination of stylet sheaths in plant tissues frequently shows little contact with phloem tissue by xylem feeders (see references above), and vice versa, with exceptions noted below. In addition, there are dozens of plant diseases whose agents are confined to the phloem, and several in xylem, that are vectored by homopterans. Potential vectors of these diseases have been carefully screened. In general, diseases restricted to the phloem tissue cannot be transmitted by insects that normally feed in the xylem, and xylem-restricted diseases cannot be vectored by insects that characteristically feed in the phloem (95, 116). This is not simply the result of vector specificity of the disease agents. For example, Pierce's disease of grapes is transmitted by 23 species of leafhoppers and three species of spittlebugs, all of which feed in the xylem, where the causal bacterium occurs. Similarly, the California aster yellows disease, which is confined to the phloem, is transmitted by 24 species of leafhoppers that feed in the phloem, but none that feed in the xylem (116). Thus, evidence provided by feeding tracks and vector specificities is strong that phloem and xylem feeders often have fastidious diets.

Many families and subfamilies of Homoptera feed characteristically in only xylem, phloem, or mesophyll. For this reason, we conclude that the morphological and chemical differences between these tissues are sufficiently great that (1) insect species adapted to feed efficiently in one cannot do so in the others, and (2) evolutionary shifts from one feeding speciality to another are rare. However, there are exceptions to both rules. First, some insect species can feed in different tissue types, either at different stages of development or concurrently. The sugarcane froghopper [*Aeneolamia varia saccharina* (Distant)] is a sequential specialist: nymphs feed on xylem, while adults specialize on the border parenchyma cells (48). Other species are true generalists in that they feed alternatively in various tissues: these include certain aphids (33, 69, 70), leafhoppers (29, 80, 115), and psyllids (119). [These examples are to be distinguished from cases of insects feeding in a variety of tissues in unsuitable or resistant hosts, while suffering reduced survival and or fecundity (13, 70, 82).] Obviously there are no absolute barriers to feeding on several kinds of plant sap, though such generalists may be less efficient feeders on any one kind of sap than are, say, xylem or phloem specialists.

In addition, one can readily find examples of evolutionary shifts in diet among tissue specialists. For example, *Empoasca fabae* (Harris), the potato leafhopper (106), and *E. flavescens* Fabr. (15) feed in phloem, although at least eight other species in the genus and virtually all other Typhlocybinae feed in mesophyll (see section on mesophyll feeders). Also, Pseudococcidae feed on either phloem or mesophyll (or both) (16, 31, 44, 107), indicating that some species have shifted their diet (and perhaps that others are in transition). Finally, biotype A of the greenbug [*Schizaphis graminum* (Rondani)] is a phloem specialist, while biotype B, which supplanted it as a major pest of some wheat varieties, feeds in the mesophyll (104).

Janzen (59) pointed out that all insect species feeding on a plant species interact with one another through their effects on the plant. This is true, provided the insects are actually feeding on the same individual plants, and the damage from their feeding is sufficiently intense. However, it does not invalidate the assignment of the insects into different feeding guilds. With some exceptions, the Homoptera and phytophagous Hemiptera can be divided into groups feeding on three distinct tissues: xylem, phloem, and mesophyll. The two vascular feeders remove water and photosynthate, respectively, while only the mesophyll feeders directly reduce the amount of photosynthetic tissue in the plant. If the insects of any group do affect their host plant significantly, then these effects will have the greatest consequences for other members of their group. This alone justifies the separation of the species on the plants into guilds. Even when feeding damage to the plants is minimal and interactions between members of a guild are unimportant, it is still a useful classification. It combines species sharing a wide range of morphological, physiological, and behavioral adaptations to the utilization of similar food, much as the classification of vertebrate herbivores into ruminants and nonruminants; these species might be expected to share other aspects of their ecologies as well.

Vascular Prokaryotes: Guild Members?

Certain plant pathogens pose intriguing problems for development of a unified guild concept, and deserve comment in a chapter with significance for disease transmission. The feeding strategies utilized by mesophyll-, phloem-, and xylem-feeding homopterans have entirely different consequences for transmission of plant disease. Mesophyll feeders are not known to transmit plant disease, presumably because their target cells are destroyed by the feeding process. Homopterans that feed in phloem and xylem, on the other hand, transmit an array of plant pathogens, including many organisms that induce diseases of serious economic importance (116). These pathogens utilize the same general resources as the insects that feed in vascular tissue: should they be elected to guild membership?

The assemblage of xylem-inhabiting prokaryotes is diverse, and includes extreme cases of generalist and specialist organisms (1). At one end of the spectrum, fungal or bacterial vascular wilt pathogcns gain acccss to tracheal elements through wounds, including those made by insects. The fungi (e.g., *Corynebacterium, Fusarium, Verticillium*) or bacteria (e.g., *Xanthomonas, Pseudomonas, Erwinia*), after entering the xylem, produce pectinases and cellulases that degrade the tracheal walls. Materials released by this process then are transported in the vessels, and eventually obstruct the vessel pores and block the transpiration stream, leading to wilt of the plant. When the organisms break through the walls they are also able to multiply in parenchymal tissue. Thus, these generalists truly overlap vascular and parenchymal tissue in their feeding strategies.

More interesting and germane to a discussion of the guilds of xylem specialists are those organisms that do not destroy the xylem tissue once they have entered it. These are the bacteria that cause such diseases as ratoon stunt of sugarcane (109) or Pierce's disease of grapevines (51). Although these pathogens attract attention as a result of the extensive damage that they cause on crops, their natural maintenance cycles are not well understood. In the case of Pierce's disease, the fastidious causative bacterium is evidently widely distributed in a variety of plant hosts (51), and is transmitted by a wide variety of cicadelline leafhoppers, which are commonly regarded to be xylem feeders (96, 118). These pathogens, unlike the generalist Gram-negative bacterial pathogens, do not multiply in conventional media, and specialized media are required for their cultivation (25, 26).

In general, the specialized media that have been successful for xylem pathogens contain much lower concentrations of nutrients than media used for cultivation of phloem pathogens. Failure of more complex media seem to be ascribable to oversupplies of nutrients, rather than failure to supply critical, limiting substances. Thus, the principal adaptive basis for utilization of xylem appears to involve interconversion of scarce nutrients from one form to another, or to utilize nutrients present in critically limiting quantities.

The array of pathogens inhabiting the phloem is also diverse. Among phloem-inhabiting prokaryotes (102) may be included spiroplasmas (helical, wall-less motile prokaryotes) (129) and a loosely defined group of noncultivable "yellows" agents (68) that are recognized and differentiated on the basis of specific symptomatology. Also included is a taxon of small fastidious bacteria, all of which have so far escaped efforts at cultivation (63, 135). The fact that yellows agents and fastidious phloem-limited bacteria resist cultivation, whereas spiroplasmas can be cultivated in a variety of media, emphasizes the nutritionally diverse abilities of different taxa that apparently utilize the same habitat (namely phloem). With phloem pathogens, it has been more difficult to discern nutritional overloads. Rather, the common approach to cultivation has been to supply as many nutrients as possible (130). Chemically defined media have been designed for several spiroplasmas of insect habitats (18). Components that have proved to be essential for spiroplasmas include amino acids, carbohydrates, sterols, vitamins, and nucleic acid precursors; it is likely that all these substances normally occur in sieve elements inhabited by the spiroplasmas.

Discovery of a diverse assortment of spiroplasma species in association with insects (22) makes it possible to assess the habitats of these microorganisms in relation to their nutritional specialization. The 25-odd cultivable spiroplasma species now under study exhibit a bewildering array of nutritional capabilities. For example, although pathways for arginine catabolism are often present, some species lack this pathway. Similarly, the degree of dependence on glucose or other fermentative pathways differs profoundly among isolates and between species. Chang and Chen (18) felt that the different requirements they observed were related to habitats of the organisms. If there is a single unifying factor in the ecology of known spiroplasmas, it is that all are closely linked to restricted host-defined habitats. Since the biochemistry of each host differs, in some cases profoundly, the genomes of the organisms have become modified for the realities of the habitat to which they have become adapted. Recent studies, in fact, suggest that there is a distinct correlation between the "clonality" of microbial clusters and the homogeneity of the habitat in which the organisms live (101). For example, *Mycoplasma pneumoniae,* a pathogen that is closely associated with mucosa of the human respiratory tract, occurs as a tight cluster of homogeneous strains (101). In contrast, *Acholeplasma* species of the *granularum* cluster, which tend to be isolated from a variety of habitats, are not so tightly clustered. *Spiroplasma* species that have proved difficult or impossible to cultivate always seem to be in special habitats. For example, the sex-ratio spiroplasmas of *Drosophila* that are vertically transmitted through the egg and selectively eliminate male progeny steadfastly resisted cultivation (134). Thus, in a single genus of microorganisms, there appears to exist a range of nutritional adaptations that range from generalist to mutualist. This is not to say that other important features of life history strategies are unimportant to these microorganisms. The requirement to find a means for dispersal to new habitats remains equally imperative. In this regard, the homopteran strategists that share the vascular tissues with the spiroplasmal and mycoplasmal pathogens of course play a central role.

Once transmission has occurred, the pathogens enter the phloem stream of the plant and induce systemic infection. There is no evidence that the phloem tissue of fresh plants chosen by the insects has a significant microbial flora. Rather, the current picture of plant vascular tissue is one of a tissue that is, despite the lack of an immune system with specific memory, normally as aseptic as the vascular tissue of animals. Presumably, the tissue is accessible only to organisms that can be introduced by the specific feeding of phloem-specialist insects.

Inoculation of vascular plants is not a rare event in nature; widely separated, occasional transmission events would probably escape notice. In fact, epidemics such as that associated with the destruction of tall coconut in Jamaica and south Florida by the lethal yellowing mollicute have forcefully called such transmission to attention (117). In certain instances of disease transmission, multiple infections have been noted (112). For example, the infection of phloem tissue by the aster yellows pathogen and oat blue dwarf virus (7) has been demonstrated, and in California, simultaneous infection of hosts by *Spiroplasma citri* and a noncultivable virescence agent, or by *S. citri* and the mycoplasmalike organisms (MLO) causing either pear decline (99) or Western X-disease of stone fruits (62), led to serious confusion concerning disease etiology before the dual nature of the infections was realized. In the neotropics, it is common to find corn plants that may be simultaneously infected by the corn stunt spiroplasma, the maize bushy stunt MLO, and the maize rayado fino virus (81). In each of these cases, the organisms may not only coinfect their plant hosts, but are transmitted by the same leafhopper species [*Dalbulus maidis* (DeLong and Wolcott)] and in some cases by the same individual insect.

The coexistence of different pathogens in phloem presents no conceptual problem to competition theory. The phloem tissue evolves, in the life of a single cell, from an actively living cell fully capable of supporting viral multiplication, to a probably lifeless element, which then serves as a conduit for photosynthate transport. One can hypothesize that a single phloem cell could, in its lifetime, support first a nondestructive virus, then a fastidious mollicute, and, finally, a more nutritionally general spiroplasma that utilizes the photosynthate of the plant rather than living protoplasm.

It is unlikely that insect parasites feeding on phloem tissue could subdivide its resources as finely as the plant pathogens. Thus, although insects, viruses, and prokaryotic microorganisms utilize the same plant tissue, we argue that the grain of exploitation, both spatial and temporal, is sufficiently different that insects and microorganisms cannot be thought of as members of the same guild; instead, the microorganisms themselves might be subdivided into guilds that depend on the stage in the life of individual phloem elements. Thus, viruses, spiroplasmas, and mycoplasmalike organisms may well comprise guilds of their own. Indeed, experiments in which competitive challenges have been deliberately induced (96) tend to support the concept that organisms of similar taxa may in fact be subject to general laws governing competition between organisms.

Summary

We reviewed the literature on Homoptera feeding to make the following points: (1) Homoptera typically feed either on phloem or xylem sap, or on the cytoplasm of mesophyll and other cells, though there are exceptions that feed more generally. (2) Each of these three feeding modes requires special morphological, physiological and behavioral adaptations, which may preclude feeding on other plant, let alone animal, tissues. (3) As a result, the food eaten by these insects is a conservative character in evolution, and often characterizes entire subfamilies or families. (4) Other features of the biology of the Homoptera, such as patterns of host specificity and their role as vectors of plant pathogens, may also be understood in light of their feeding strategies. For reasons 1–4, it is useful to categorize Homoptera, where possible, into three distinct feeding guilds. (5) Plant pathogens such as bacteria, fungi, viruses, and spiroplasmas that inhabit the vascular tissues also tend to be tissue specialists, though they can subdivide these tissues on a much finer spatial and temporal scale. As a result, the pathogens and the Homoptera that vector them are best considered members of different, only partially overlapping guilds.

Acknowledgments. We thank the following people for their careful reading and helpful criticism of this paper: M.F. Claridge, H.S. Horn, L.R. Nault, A.H. Purcell, D.E. Ullman, and N. Waloff.

References

1. Agrios, G.N., 1978, *Plant Pathology,* Academic Press, New York. 703 pp.
2. Aoki, S., 1978, Two pemphigids with first instar larvae attacking predatory intruders (Homptera, Aphidoidea), *New Entomol.* **27:**7–12.
3. Arnaud, G., 1918, Les asterinees, *Anu. EC. Agric. Montpellier,* **1918:**52–65.

4. Backus, E.A., and McLean, D.L., 1982, The sensory systems and feeding behavior of leafhoppers. I. The aster leafhopper, *Macrosteles fascifrons* Stål (Homoptera: Cicadellidae), *J. Morphol.* **172:**359–378.
5. Backus, E.A., and McLean, D.L., 1983, The sensory systems and feeding behavior of leafhoppers. II. A comparison of the sensillar morphologies of several species (Homoptera: Cicadellidae), *J. Morphol.* **176:**3–14.
6. Backus, E.A., and McLean, D.L., 1985, Behavioral evidence that the precibarial sensilla of leafhoppers are chemosensory and function in host discrimination, *Entomol. Exp. Appl.* **37:**219–228.
7. Banttari, E.E., and Zeyen, R.J., 1979, Interactions of mycoplasmalike organisms and viruses in dually infected leafhoppers, planthoppers and plants, in: K.F. Harris and K. Maramorosch (eds.), *Leafhopper Vectors and Plant Disease Agents,* Academic Press, New York, pp. 327–347.
8. Bennet, C.W., 1934, Plant–tissue relations of the sugar-beet curly-top virus, *J. Agric. Res.* **48:**665–701.
9. Bernard, J., Pinet, J.M., and Boistel, J., 1970, Electrophysiologie des récepteurs des stylets maxillaires de *Triatoma infestans*—Action de la température et de la teneur en eau de l'air, *J. Insect Physiol.* **16:**2157–2180.
10. Blackmore, S., 1981, Penetration of the host plant tissues by the stylets of the coccoid *Icerya seychellarum* (Coccoidea: Margaroidea), *Atoll Res. Bull.* **255:**33–38.
11. Bollard, E.G., 1960, Transport in the xylem, *Annu. Rev. Plant Physiol.* **11:**141–166.
12. Byers, R.A., and Wells, H.D., 1966, Phytotoxemia of coastal bermudagrass caused by the two-lined spittlebug, *Prosapia bicincta* (Homoptera: Cercopidae), *Ann. Entomol. Soc. Am.* **59:**1067–1071.
13. Campbell, B.C., McLean, D.L., Kinsey, M.G., Jones, K.C., and Dreyer, D.L., 1982, Probing behavior of the greenbug (*Schizaphis graminum,* biotype C) on resistant and susceptible varieties of sorghum, *Entomol. Exp. Appl.* **31:**140–146.
14. Capoor, S.P., 1949, Feeding methods of the white-fly, *Curr. Sci.* **18:**82–83.
15. Carle, P., and Moutous, G., 1965, Observations sur le mode de nutrition sur vigne de quatre éspèces de cicadelles, *Ann. Epiphyt.* **16:**333–354.
16. Carter, W., 1960, *Phenococcus solani* Ferris, a toxigenic insect, *J. Econ. Entomol.* **53:**322–323.
17. Carter, W., 1973, *Insects in Relation to Plant Diseases,* 2nd ed., Wiley, New York. 759 pp.
18. Chang, C.J., and Chen, T.A., 1982, Spiroplasmas: Cultivation in chemically defined medium, *Science* **215:**1121–1122.
19. Cheung, W.W.K., and Marshall, A.T., 1973, Water and ion regulation in cicadas in relation to xylem feeding, *J. Insect. Physiol.* **19:**1801–1816.
20. Claridge, M.F., and Wilson, M.R., 1976, Diversity and distribution patterns of some mesophyll-feeding leafhoppers of temperate woodland canopy, *Ecol. Entomol.* **1:**231–250.
21. Claridge, M.F., and Wilson, M.R., 1981, Host plant associations, diversity and species–area relationships of trees and shrubs in Britain, *Ecol. Entomol.* **6:**217–238.
22. Clark, T.B., 1982, Spiroplasmas: Diversity of arthropod reservoirs and host–parasite relationships, *Science* **217:**57–59.

23. Cobben, R.H., 1978, *Evolutionary Trends in Heteroptera. Part II. Mouthpart-Structures and Feeding Strategies,* H. Veenman and Zonen B.V., Wageningen, the Netherlands.
24. Dadd, R.H., 1973, Insect nutrition: current developments and metabolic implications, *Annu. Rev. Entomol.* **18:**381–420.
25. Davis, M.J., Purcell, A.H., and Thomson, S.V., 1978, Pierce's disease of grapevines: Isolation of the causal bacterium, *Science* **199:**75–77.
26. Davis, M.J., Gillaspie, A.G., Harris, R.W., and Lawson, R.H., 1980, Ratoon stunting disease of sugarcane: Isolation of the causal bacterium, *Science* **210:**1365–1367.
27. Day, M.F., and McKinnon, A., 1951, A study of some aspects of the feeding of the jassid *Orosius, Aust. J. Sci. Res. B* **4:**125–135.
28. Day, M.F., and Irzykiewicz, H., 1953, Feeding behaviour of the aphids *Myzus persicae* and *Brevicoryne brassicae* studied with radiophosphorus, *Aust. J. Biol. Sci.* **6:**98–108.
29. Day, M.F., Irzykiewizc, H., and McKinnon, A., 1952, Observations on the feeding of the virus vector *Orosius argentatus* (Evans) and comparisons with certain other jassids, *Aust. J. Sci. Res. B* **5:**128–142.
30. Eastop, V.F., 1973, Deductions from the present day host plants of aphids and related insects, in: H.F. van Emden (ed.), *Insect/Plant Relationships,* Wiley, New York, pp. 157–178.
31. Entwistle, P.F., and Longworth, J.F., 1963, The relationships between cacao viruses and their vectors: The feeding behaviours of three mealybug (Homoptera: Pseudococcidae) species, *Ann. Appl. Biol.* **52:**387–391.
32. Esau, K., 1961, *Plants, Viruses, and Insects,* Harvard University Press, Cambridge, Massachusetts. 110 pp.
33. Esau, K., Namba, R., and Rasa, E.A., 1961, Studies on penetration of sugar beet leaves by stylets of *Myzus persicae, Hilgardia* **30:**517–529.
34. Eyer, J.R., 1937, Physiology of psyllid yellows of potatoes, *J. Econ. Entomol.* **30:**891–898.
35. Eyer, J.R., and Crawford, R.F., 1933, Observations on the feeding habits of the potato psyllid (*Paratrioza cockerelli* Sulc.) and the pathological history of the "psyllid yellows" which it produces, *J. Econ. Entomol.* **26:**846–850.
36. Fife, J.M., and Frampton, V.L., 1936, The pH gradient extending from the phloem into the parenchyma of the sugar beet and its relation to the feeding behavior of *Eutettix tenellus, J. Agric. Res.* **53:**581–593.
37. Fife, J.M., Price, C., and Fife, D.C., 1962, Some properties of phloem exudate collected from roots of sugar beet, *Plant Physiol.* **37:**791–792.
38. Fisher, J.B., and Tsai, J.H., 1977, Feeding sites of leafhoppers and planthoppers on plant tissues in: *Proceedings Third International Council on Lethal Yellowing,* p. 23.
39. Forbes, A.R., 1972, Innervation of the stylets of the pear psylla, *Psylla pyricola* (Homoptera: Aleyrodidae), *J. Entomol. Soc. Br. Columbia* **69:**27–30.
40. Forbes, A.R., 1977, The mouthparts and feeding mechanisms of aphids, in: K.F. Harris and K. Maramorosch (eds.), *Aphids as Virus Vectors,* Academic Press, New York, pp. 83–103.
41. Forbes, A.R., and MacCarthy, H.R., 1969, Morphology of the Homoptera, with emphasis on virus vectors, in: K. Maramorosch (ed.), *Viruses, Vectors and Vegetation,* Intersciences, New York, pp. 211–234.

42. Forbes, A.R., and Raine, J., 1973, The stylets of the six-spotted leafhopper, *Macrosteles fascifrons* (Homoptera: Cicadellidae), *Can. Entomol.* **105:**559–567.
43. Foster, S., 1981, Sensory receptors in the mouthparts of the rice brown planthopper *Nilaparvata lugens* (Stål), *Acta Entomol. Fenn.* **38:**23–24.
44. Glass, E.H., 1944, Feeding habits of two mealybugs, *Pseudococcus comstockii* (Kuw.) and *Phenacoccus colemani* (Ehr), *Tech. Bull. Va. Agric. Exp. Stn.* **95:**1–16.
45. Goodchild, A.J.P., 1966, Evolution of the alimentary canal in the Hemiptera, *Biol. Rev.* **41:**97–140.
46. Gunthart, H., and Gunthart, M., 1981, Biology and feeding behaviour of *Aguriahana germari* (Zett.) (Homoptera, Auchenorrhyncha, Typhlocybinae), *Acta Entomol. Fenn.* **38:**24.
47. Gunthart, M.S., and Wanner, H., 1981, The feeding behaviour of two leafhoppers on *Vicia faba, Ecol. Entomol.* **6:**17–22.
48. Hagley, E.A., and Blackman, J.A., 1966, Site of feeding of the sugarcane froghopper, *Aeneolamia varia saccharina* (Homoptera: Cercopidae), *Ann. Entomol. Soc. Am.* **59:**1289–1291.
49. Halkka, O., Raatikainen, M., Vasarainen, A., and Heinoven, L., 1967, Ecology and ecological genetics of *Philaenus spumarius* (L.) (Homoptera), *Ann. Zool. Fenn.* **4:**1–18.
50. Hargreaves, E., 1915, The life history and habits of the greenhouse whitefly (*Aleyrodes vaporariorum* Westd.), *Ann. Appl. Biol.* **1:**303–334.
51. Hewitt, W.B., 1970, Pierce's disease of *Vitis* species, in: N.W. Frazier (ed.), *Virus Diseases of Small Fruits and Grapevines,* University of California Press, Berkeley, California, pp. 196–200.
52. Hildebrand, E.M., 1961, Relations between whitefly and sweetpotato tissue in transmission of yellow dwarf virus, *Science* **133:**282–284.
53. Hodkinson, I.D., 1973, The biology of *Strophingia ericae* (Curtis) (Homoptera, Psylloidea) with notes on its primary parasite *Tetrastichus actis* (Walker) (Hum., Eulophidae), *Norsk. Entomol. Tidsskr.* **20:**237–243.
54. Hodkinson, I.D., 1974, The biology of the Psylloidea (Homoptera): A review, *Bull. Entomol. Res.* **64:**325–339.
55. Horne, A.S., and Lefroy, H.M., 1915, Effects produced by sucking insects and red spider upon potato foliage, *Ann. Appl. Biol.* **1:**370–386.
56. Horsfield, D., 1977, Relationship between feeding of *Philaenus spumarius* (L.) and the amino acid concentration in the xylem sap, *Ecol. Entomol.* **2:**259–266.
57. Horsfield, D., 1978, Evidence of xylem feeding by *Philaenus spumarius* (L.) (Homoptera: Cercopidae), *Entomol. Exp. Appl.* **24:**95–99.
58. Houston, B.R., Esau, K., and Hewitt, W.B., 1947, The mode of feeding and the tissues involved in the transmission of Pierce's disease virus in grape and alfalfa, *Phytopathology* **37:**247–253.
59. Janzen, D.H., 1973, Host plants as islands, II. Competition in evolutionary and contemporary time, *Am. Nat.* **107:**786–790.
60. King, W.V., and Cook, W.S., 1932, Feeding punctures of mirids and other plant-sucking insects and their effect on cotton, *U.S. Dep. Agric. Tech. Bull.* **296:**1–11.
61. Kinsey, M.G., and McLean, D.L., 1967, Additional evidence that aphids ingest through an open stylet sheath, *Ann. Entomol. Soc. Am.* **60:**1263–1265.

62. Kloepper, J.W., and Garrott, D.G., 1983, Evidence for a mixed infection of spiroplasmas and nonhelical mycoplasmalike organisms in cherry with X-disease, *Phytopathology* **73:**357–360.
63. Laflèche, D., and Bové, J.M., 1970, Structures de type mycoplasme dans les feuilles d'orangers atteints de la maladie du "Greening", *C.R. Hebd. Acad. Sci. Paris. D* **270:**1915–1917.
64. Llewellyn, M., 1982, The energy economy of fluid-feeding herbivorous insects, in: J.H. Visser and A.K. Minks (eds.), *Proceedings 5th International Symposium on Insect–Plant Relationships,* Pudoc, Wageningen, the Netherlands, pp. 243–251.
65. Marshall, C., and Sagar, G.R., 1976, Transport in the phloem, in: M.A. Hall (ed.), *Plant Structure, Function and Adaptation,* Macmillan, London, pp. 254–293.
66. Mattson, W.J., 1980, Herbivory in relation to plant nitrogen content, *Annu. Rev. Ecol. Syst.* **11:**119–161.
67. McClure, M.S., and Price, P.W., 1976, Ecotope characteristics of coexisting *Erythroneura* leafhoppers (Homoptera: Cicadellidae) on sycamore, *Ecology* **57:**928–940.
68. McCoy, R.E., 1979, Mycoplasmas and yellows diseases, in: R.F. Whitcomb and J.G. Tully (eds.), *The Mycoplasmas,* Academic Press, New York, pp. 229–264.
69. McLean, D.L., and Kinsey, M.G., 1967, Probing behavior of the pea aphid, *Acyrthosiphon pisum.* I. Definitive correlation of electronically recorded waveforms with aphid probing activities, *Ann. Entomol. Soc. Am.* **60:**400–406.
70. McLean, D.L., and Kinsey, M.G., 1968, Probing behavior of the pea aphid, *Acyrthosiphon pisum.* II. Comparisons of salivation and ingestion in host and non-host plant leaves, *Ann. Entomol. Soc. Am.* **61:**730–739.
71. McLean, D.L., and Kinsey, M.G., 1984, The precibarial valve and its role in the feeding behavior of the pea aphid, *Acyrthosiphon pisum, Bull. Entomol. Soc. Am.* **30:**26–31.
72. McMurtry, J.A., and Stanford, E.H., 1960, Observations of feeding habits of the spotted alfalfa aphid on resistant and susceptible alfalfa plants, *J. Econ. Entomol.* **53:**714–717.
73. McNeil, S., and Southwood, T.R.E., 1978, The role of nitrogen in the development of insect/plant relationships, in: J.B. Harborne (ed.), *Biochemical Aspects of Plant and Animal Coevolution,* Academic Press, London, pp. 77–98.
74. Miles, P.W., 1968, Insect secretions in plants, *Annu. Rev. Phytopathol.* **6:**137–164.
75. Miles, P.W., McLean, D.L., and Kinsey, M.G., 1964, Evidence that two species of aphid ingest food through an open stylet sheath, *Experientia* **20:** 582.
76. Miller, N.C.E., 1971, *The Biology of the Heteroptera,* 2nd ed., E.W. Classey, Hampton, England. 206 pp.
77. Mittler, T.E., 1957, Studies on the feeding and nutrition of *Tuberolachnus salignus* (Gmelin) (Homoptera, Aphididae). I. The uptake of phloem sap, *J. Exp. Biol.* **34:**334–341.
78. Mittler, T.E., 1958, Studies on the feeding and nutrition of *Tuberolachnus salignus* (Gmelin) (Homoptera: Aphididae). II. The nitrogen and sugar com-

position of ingested phloem sap and excreted honeydew, *J. Exp. Biol.* **35**:74–84.
79. Mittler, T.E., 1967, Water tensions in plants—An entomological approach, *Ann. Entomol. Soc. Am.* **60**:1074–1076.
80. Naito, A., 1977, Feeding habits of leafhoppers, *Jpn. Agric. Res. Q.* **11**:115–119.
81. Nault, L.R., and Bradfute, O.E., 1979, Corn stunt: Involvement of a complex of leafhopper-borne pathogens, in: K. Maramorosch and K.F. Harris (eds.), *Leafhopper Vectors and Plant Disease Agents,* Academic Press, New York, pp. 561–586.
82. Nault, L.R., and Styer, W.E., 1972, Effects of sinigrin on host selection by aphids, *Entomol. Exp. Appl.* **15**:423–437.
83. Newby, R., 1979, Growth and feeding in two species of Machaerotidae (Homoptera), *Aust. J. Zool.* **27**:395–401.
84. Nielson, M.W., 1979, Taxonomic relationships of leafhopper vectors of plant pathogens, in: K. Maramorosch and K.F. Harris (eds.), *Leafhopper Vectors and Plant Disease Agents,* Academic Press, New York, pp. 3–27.
85. Owen, D.F., 1978, Why do aphids synthesize melezitose?, *Oikos,* **31**:264–267.
86. Parrish, W.B., 1967, The origin, morphology, and innervation of aphid stylets (Homoptera), *Ann. Entomol. Soc. Am.* **60**:273–276.
87. Pate, J.S., 1973, Uptake, assimilation and transport of nitrogen compounds by plants, *Soil Biol. Biochem.* **5**:109–119.
88. Pate, J.S., 1976, Nutrients and metabolites of fluids recovered from xylem and phloem: Significance in relation to long distance transport in plants, in: I.F. Wardlaw and J.B. Passiours (eds.), *Transport and Transfer Processes in Plants,* Academic Press, New York, pp. 253–281.
89. Pinet, J.M., and Bernard, J., 1972, Essai d'interprétation du mode d'action de la vapeur d'eau et de la température sur un récepteur d'insecte, *Ann. Zool. Ecol. Anim.* **4**:483–495.
90. Pollard, D.G., 1955, Feeding habits of the cotton whitefly, *Bemesia tabaci* Genn. (Homoptera: Aleyrodidae), *Ann. Appl. Biol.* **43**:664–671.
91. Pollard, D.G., 1968, Stylet penetration and feeding damage of *Eupteryx melissae* Curtis (Hemiptera: Cicadellidae) on sage, *Bull. Entomol. Res.* **58**:55–71.
92. Pollard, D.G., 1973, Plant penetration by feeding aphids (Hemiptera, Aphidoidea): A review, *Bull. Entomol. Res.* **62**:631–714.
93. Prestidge, R.A., 1982, The influence of nitrogenous fertilizer on the grassland Auchenorrhyncha (Homoptera), *J. Appl. Ecol.* **19**:735–749.
94. Prestidge, R.A., and McNeill, S., 1981, The role of nitrogen in the ecology of grassland Auchenorrhyncha, in: J.A. Lee, S. McNeill, and I.H. Rorison (eds.), *Nitrogen as an Ecological Factor,* Blackwell, Oxford, pp. 257–281.
95. Purcell, A.H., 1979, Leafhopper vectors of xylem-borne plant pathogens, in: K. Maramorosch and K.F. Harris (eds.), *Leafhopper Vectors and Plant Disease Agents,* Academic Press, New York, pp. 603–627.
96. Purcell, A.H., 1982, Evolution of the insect vector relationship, in: M.S. Mount and G.H. Lacey (eds.), *Phytopathogenic Prokaryotes,* Vol. I, Academic Press, New York, pp. 121–156.
97. Putnam, W.L., 1941, The feeding habits of certain leafhoppers, *Can. Entomol.* **73**:39–53.

98. Putnam, W.L., 1953, Notes on the bionomics of some Ontario cercopids (Homoptera), *Can. Entomol.* **85**:244–248.
99. Raju, B.C., Nyland, G., and Purcell, A.H., 1983, Current status of the etiology of pear decline, *Phytopathology* **73**:350–353.
100. Raven, J.A., 1984, Phytophages of xylem and phloem: A comparison of animal and plant sap-feeders, *Adv. Ecol. Res.* **13**:135–234.
101. Razin, S., Tully, J.G., Rose, D.L., and Barile, M.F., 1983, DNA cleavage patterns as indicators of genotypic heterogeneity among strains of *Acholeplasma* and *Mycoplasma* species, *J. Gen. Microbiol.* **129**:1935–1944.
102. Saglio, P.H.M., and Whitcomb, R.F., 1979, Diversity of wall-less prokaryotes in plant vascular tissue, fungi, and invertebrate animals, in: R.F. Whitcomb and J.G. Tully (eds.), *The Mycoplasmas,* Vol. III, Academic Press, New York, pp. 1–36.
103. Saxena, K.N., 1954, Feeding habits and physiology of digestion of certain leafhoppers (Homoptera: Jassidae), *Experientia* **10**:383–384.
104. Saxena, P.N., and Chada, H.L., 1971, The greenbug, *Schizaphis graminum.* I. Mouth parts and feeding habits, *Ann. Entomol. Soc. Am.* **64**:897–904.
105. Schetters, C., 1960, Untersuchungen uber die art und die folgewirkung des saugvorganges bei der San Jose-schillaus (*Aspidiotus perniciosus* Comst.), *Z. Angew. Entomol.* **46**:277–322.
106. Smith, F.F., and Poos, F.W., 1931, The feeding habits of some leafhoppers of the genus *Empoasca, J. Agric. Res.* **43**:267–286.
107. Smith, K.M., 1926, A comparative study of the feeding methods of certain Hemiptera and of the resulting effects upon the plant tissue, with special reference to the potato plant, *Ann. Appl. Biol.* **13**:109–139.
108. Sogawa, K., 1973, Feeding of the rice plant- and leafhoppers, *Rev. Plant Protect. Res.* **6**:31–43.
109. Steindl, D.R., 1961, Ratoon stunting, in: J.P. Martin, E.V. Abbott, and C.G. Hughes (eds.), *Sugarcane Diseases of the World,* Elsevier, Amsterdam.
110. Stiling, P.D., 1980, Competition and coexistence among *Eupteryx* leafhoppers (Hemiptera: Cicadellidae) occurring on stinging nettle (*Urtica dioica* L.), *J. Anim. Ecol.* **49**:793–805.
111. Storey, H.H., 1938, Investigations of the mechanisms of the transmission of plant viruses by insect vectors. II. The part played by puncture in transmission, *Proc. R. Soc. B (Lond.)* **125**:455–477.
112. Sylvester,. E., 1975, Multiple acquisition of viruses and vector-dependent prokaryotes: Consequences of transmission, *Annu. Rev. Entomol.* **30**:71–88.
113. Tonkyn, D.W., 1985a, Temporal and spatial convergence in a guild of sap-feeding insects on Early Goldenrod (*Solidago juncea* Ait; Asteraceae), Ph.D thesis, Princeton University, Princeton, New Jersey. 160 pp.
114. Tonkyn, D.W., 1985b, Predator-mediated mutualism: Theory and tests in the Homoptera, *J. Theor. Biol.* **118**:15–31.
115. Triplehorn, B.W., Nault, L.R., and Horn, D.J., 1984, Feeding behavior of *Graminella nigrifons* (Forbes), *Ann. Entomol. Soc. Am.* **77**:102–107.
116. Tsai, J.H., 1979, Vector transmission of mycoplasmal agents of plant diseases, in: R.F. Whitcomb and J.G. Tully (eds.), *Plant and Insect Mycoplasmas,* Academic Press, New York, pp. 265–397.
117. Tsai, J.H., 1981, Transmission of lethal yellowing mycoplasma by *Myndus crudus,* in: K. Maramorosch and S.P. Raychaudhuri (eds.), *Mycoplasma Diseases of Trees and Shrubs,* Academic Press, New York, pp. 211–229.

118. Turner, W.F., and Pollard H.N., 1959, Life histories and behavior of five insect vectors of phony peach disease, *U.S. Dep. Agric. Tech. Bull.* **1188:**1–57.
119. Ullman, D.E., 1985, The sensory systems and feeding behavior of the pear psylla, *Psylla pyricola* Foerster (Homoptera: Psyllidae), Ph.D. thesis, University of California, Davis. 106 pp.
120. Van Emden, H.F., 1972, Aphids as phytochemists, in: J.B. Harborne (ed.), *Phytochemical Ecology,* Academic Press, New York, pp. 25–43.
121. Van Emden, H.F., Eastop, V.F., Hughes, R.D., and Way, M.J., 1969, The ecology of *Myzus persicae, Annu. Rev. Entomol.* **14:**197–270.
122. Varma, P.M., 1963, Transmission of plant viruses by whiteflies, *Natl. Inst. Sci. India Bull.* **24:**11–33.
123. Waloff, N., 1980, Studies on grassland leafhoppers (Auchenorrhyncha, Homoptera) and their natural enemies, *Adv. Ecol. Res.* **11:**81–215.
124. Wardlaw, I.F., 1968, The control and pattern of movement of carbohydrates in plants, *Bot. Rev.* **34:**79–105.
125. Way, M.J., 1963, Mutualism between ants and honeydew producing Homoptera, *Annu. Rev. Entomol.* **8:**307–344.
126. Webb, J.W., 1977, The life history and population dynamics of *Acizzia russellae* (Homoptera: Psyllidae), *J. Entomol. Soc. S. Afr.* **40:**37–46.
127. Wensler, R.J.D., 1974, Sensory innervation monitoring movement and position in the mandibular stylets of the aphid, *Brevicoryne brassicae, J. Morphol.* **143:**349–364.
128. Wensler, R.J., and Filshie, B.K., 1969, Gustatory sense organs in the food canal of aphids, *J. Morphol.* **129:**473–492.
129. Whitcomb, R.F., 1981, The biology of spiroplasmas, *Annu. Rev. Entomol.* **26:**397–425.
130. Whitcomb, R.F., 1982, Culture media for spiroplasmas, in: S. Razin and J.G. Tully (eds.), *Methods in Mycoplasmology,* Vol. I, Academic Press, New York, pp. 147–158.
131. White, J., and Strehl, C.E., 1978, Xylem feeding by periodical cicada nymphs on tree roots, *Ecol. Entomol.* **3:**323–327.
132. White, T.C.R., 1976, Weather, food and plagues of locusts, *Oecologia* **22:**119–134.
133. Wiegart, R.G., 1964, The ingestion of xylem sap by meadow spittlebugs, *Philaenus spumarius* (L.), *Am. Midl. Nat.* **71:**422–428.
134. Williamson, D.L., Steiner, T., and McHarrity, G.J., 1983, Spiroplasma taxonomy and identification of the sex ratio organisms: Can they be cultivated?, *Yale J. Biol. Med.* **56:**583–592.
135. Windsor, I.M., and Black, L.M., 1973, Evidence that clover club leaf is caused by a rickettsia-like organism, *Phytopathology* **63:**1139–1148.
136. Woodburn, T.L., and Lewis, E.E., 1973, A comparative histological study of the effects of feeding by nymphs of four psyllid species on the leaves of eucalypts, *J. Aust. Entomol. Soc.* **12:**134–138.
137. Zimmerman, M.H., 1960, Transport in the phloem, *Annu. Rev. Plant Physiol.* **11:**167–190.

Index